Participatory Redevelopment as Contested Arenas in China

Xiang Li

Participatory Redevelopment as Contested Arenas in China

Power Relations and Institutional Transformation

Springer

Xiang Li
Department of Urban Planning
Xiamen University
Xiamen, Fujian, China

ISBN 978-981-95-9393-4 ISBN 978-981-95-9394-1 (eBook)
https://doi.org/10.1007/978-981-95-9394-1

This Springer imprint is published by the registered company Springer Nature Singapore Pte Ltd.
The registered company address is: 152 Beach Road, #21-01/04 Gateway East, Singapore 189721, Singapore

Preface

Urban redevelopment has become a defining feature of contemporary urban transformation, particularly in rapidly urbanizing contexts such as China. Over recent decades, redevelopment has increasingly been framed not only as a technical or economic intervention, but also as a governance reform, with participatory and collaborative mechanisms promoted as means to reconcile economic growth, social stability and urban justice. Yet at the heart of these participatory reforms lie complex power dynamics—shaping who gets to participate, whose interests are prioritized, and how decisions are legitimized. Despite the expanding body of scholarship on power relations and participation in urban redevelopment, important questions remain concerning how power is configured, exercised and transformed through participatory reforms.

Much existing research on urban power relations has offered valuable insights into elite influence, coalition-building and decision-making. However, power dynamics are often examined primarily through the behaviours and strategies of individual actors, with limited attention to the broader institutional environments within which these interactions unfold. At the same time, while influential theories have highlighted the embeddedness of power in structures and practices, their high level of abstraction and the lack of operational analytical tools have constrained systematic empirical analysis. As a result, existing approaches have struggled to capture the reciprocal and dynamic relationship between institutional transformation and localized power interactions in concrete redevelopment settings.

These limitations are particularly evident in studies of participatory redevelopment in China. Recent participatory policy reforms have frequently been treated as incremental governance innovations rather than as deeper institutional transformations that reshape redevelopment mechanisms. While such reforms introduce new participatory channels and reallocate certain bargaining capacities to non-state actors, they continue to operate within governance arrangements in which the state retains strong agenda setting authority. How participatory reforms simultaneously enable limited empowerment and reproduce, or re-centre, state dominance remains insufficiently examined. Moreover, empirical research has largely focused on specific redevelopment contexts or single cases, offering limited insight into how different participatory redevelopment modes generate divergent power configurations and justice outcomes.

This book seeks to address these gaps by developing a power arena analytical framework for participatory urban redevelopment. Rather than treating participation as a procedural ideal or institutional endpoint, it conceptualizes participation as a dynamic series of cross-sectional "relational arenas" shaped by the interaction between institutional arrangements and actors' mobilzsation of diverse power resources. Through comparative case studies of participatory redevelopment in Shenzhen, the book examines how different redevelopment modes reconfigure bargaining relationships among actors while simultaneously revealing the mechanisms through which state dominance is reconstituted under participatory governance.

By foregrounding the reciprocal relationship between institutional transformation and localized power interactions, this book contributes to broader debates on urban redevelopment. While grounded in the Chinese context, the analytical insights developed here extend beyond a single national setting and offer a lens through which participatory redevelopment and power relations can be examined in other cities undergoing rapid institutional and spatial change.

Xiamen, China Xiang Li

Acknowledgment My deepest gratitude goes first and foremost to my supervisor, Professor Sun Sheng Han. Since the beginning of my doctoral journey in 2015, Professor Han has provided unwavering guidance, steadfast support and sustained intellectual inspiration. He has played an indispensable role in shaping and refining my research ideas, guiding my academic direction, carefully reviewing my presentations and publications, and substantially improving the overall quality of this thesis. Beyond scholarly supervision, he generously offered me opportunities to serve as a teaching and research assistant. These experiences not only strengthened my pedagogical and research skills but also enriched my academic life in Melbourne.

Professor Han's support extended well beyond day-to-day supervision. He encouraged and facilitated my participation in international conferences and introduced me to valuable academic and professional networks, which have opened important pathways for my future career. I consider myself exceptionally fortunate to have been mentored by a scholar of such kindness, generosity and academic integrity. His profound knowledge, intellectual rigour and persistent pursuit of scholarly excellence have shaped not only my academic development but also my personal values. It is both a privilege and an honour to regard him as my academic mentor and, indeed, a guide in life.

I am equally grateful to my panel chair, Dr. Hao Wu, whose consistent encouragement over the years has been a vital source of confidence throughout my research journey. His insightful feedback and generous sharing of expertise greatly enhanced the quality of my work, particularly during the revision of manuscripts for publication. I also extend my sincere appreciation to Professor Kevin O'Connor and Professor Ray Green for their continued support and constructive input at every panel meeting. My heartfelt thanks further go to Professor Shenjing He of the University of Hong Kong and Professor Yong Tu at the National University of Singapore, whose valuable comments and suggestions were instrumental in refining and strengthening this dissertation.

My appreciation also goes to all those who offered help and support along the way. I am grateful to Ka Ling Cheung, Qike Wang, Yimin Cheng, Ke Song, Qianjin Zhang, Yinrui Xie, Chen Li, Yee Kee Ku, Tayyab Ahmad, Xin Yang, Yuelong Yang, Junxian Chen, Xi Ma, Hong Jiang, Fujie Rao, Jinlong Liu, Yang Zhao, Xiaoran Huang, Haifeng Zhao, Yueer He, Ziming Xie, Ada Chan, Wenjing Zhang, Haijuan Zhao, Wenyu Zhang, Lian Zhou, Fan Yi, Neeraj Dangol, Ahmed Hassem, Hamid Amouzad Khalili and Citra Ongkowijoyo, among many others. I also wish to thank my friends afar—Xin Guo, Yanshuo Liu, Fangzhou Tian, Weikai Liu, Jixiang Liu and Jidong Zhang—whose presence and encouragement made the doctoral journey less solitary and far more meaningful. I gratefully acknowledge Wenzhuo Wang for her contributions to the optimization of the figures in this book. I am especially thankful to all interviewees and local collaborators in Shenzhen, whose generosity, openness and willingness to share their time and experiences were essential to the success of my fieldwork.

Finally, and most importantly, I owe my deepest gratitude to my family. Your unconditional love, patience and unwavering belief in my aspirations have sustained me through every challenge. It is your steadfast support that gave me the strength and resilience to complete this doctoral journey.

Competing Interests The author has no competing interests to declare that are relevant to the content of this manuscript.

Contents

Chapter 1
Introduction

1.1 Research Problem and Research Gaps

1.1.1 Statement of Problem

Over the past decade, China has entered a new phase of development characterized by an ageing population, a declining birth rate, weakening domestic demand and a slowdown in urbanization and economic growth (Balding, 2019). The marginal benefit of the traditional stimulus that relied on massive investment in production factors is flagging dramatically, leading to sustained economic downward pressure (Xinhua News, 2017). On the other hand, three decades of headlong urbanization process have resulted in severe social and ecological issues, including ghost towns, waste of land resources, diminishing farmland, environmental quality degradation, voracious energy consumption and increasing local debt and so on, threatening the sustainable development of Chinese cities (Chen et al., 2016; Wu et al., 2006). These crises are rooted in a traditional growth-oriented planning paradigm that has long facilitated extensive urban expansion (Tian & Yao, 2018). Local governments, key players in campaigns for urban growth, often secure additional land quotas from higher-level governments by overestimating population growth in master plans, thereby enabling encroachment on greenfield land and contributing to extensive urban sprawl (Chen et al., 2015; Li et al., 2019). The pervasive urban crisis and a flagging urbanization process suggest that the land-expansion–driven model of economic and urban growth has reached its limits (Lin & Wu, 2015).

In response to these challenges, the Chinese government has pursued a series of supply-side structural reforms and innovation-driven strategies aimed at restructuring the economy towards a more sustainable development path (Jiang, 2019). Correspondingly, in the realm of land use planning, multilevel governments have introduced urban redevelopment policies to shift from outward urban expansion to the intensification of existing built-up areas, particularly in developed megacities facing acute land constraints (Li et al., 2019; 2024). Urban redevelopment policies

X. Li, *Participatory Redevelopment as Contested Arenas in China*,
https://doi.org/10.1007/978-981-95-9394-1_1

have thus become a key instrument for restructuring urban land use and industrial functions, marking a broader transition in China's urbanization process.

Previous studies show that the core concern regarding urban redevelopment is to effectively balance the interests of various stakeholders so as to offset resistance against forced demolition, paltry compensation and social instability (e.g. Bunker et al., 2002; Han & O'Connor, 2008; McCrea & Walters, 2012). In the Chinese context, the government-monopolized land transfer approach has resulted in unfairly redistributive resources and pervasive community mobilization, creating significant barriers to urban redevelopment (He & Wu, 2009; Shin, 2009). Besides, the property rights complexities and stakeholders' interests in urban redevelopment projects have raised substantial transaction costs (Zhao, 2014). It is thus critical and essential to pacify social resistance and ensure redistributive fairness in the implementation of urban redevelopment policies (Zhao, 2011; Zou, 2015, 2017).

To guide redistribution practice among stakeholders in redevelopment projects, the provincial government of Guangdong—one of the provinces that pioneered China's economic reform (Hayter & Han, 1998), introduced an interest-sharing mechanism into its urban redevelopment policy in 2009 (Li & Liu, 2018; Ye, 2011). In line with the provincial policy, Shenzhen, a Special Economic Zone established by the central government but administratively part of Guangdong Province, began experimenting with several localized participatory redevelopment approaches to facilitate interactions among stakeholders. Among these, community participation mechanisms have been established to secure affected residents' involvement in the decision-making process of urban redevelopment.

This institutional transformation has turned participatory urban redevelopment into a contested arena in which residents deploy their growing power to engage with traditionally powerful actors such as local governments and developers. As a result, stakeholders' configurations, interactions and power relations have been substantially reshaped, raising questions about how power is distributed among stakeholders, how emerging power dynamics influence redevelopment outcomes, and how participatory redevelopment mechanisms have evolved in the implementation of experimental policies. These are the fundamental questions this book is aspired to address.

1.1.2 Research Gaps

Existing research on power relations has certain limitations in both theoretical and practical aspects. Three theoretical gaps are identified in the literature. First, existing debates about power relations in urban (re)development examine the distribution of power among individual actors while largely operating in an institutional vacuum, thereby overlooking exogenous influences. Representative approaches include the community power debate, growth machine theory and urban regime theory. Their analyses are framed by the dichotomy of "pluralism versus elitism" and primarily emphasize who controls urban redevelopment (Dahl, 1961; Hunter, 1953; Logan & Molotch, 1987; Molotch, 1976; Stone, 1989, 1998). Consequently, these theories divert the focus of urban politics towards the implication of human agents' behaviours as endogenous influences. Power dynamics are thus conceptualized as a linear process of agenda formation, policy output and impact, with limited attention paid to their interaction with the institutional context as exogenous influences.

Second, as exemplified by Gidden's structuration theory and Foucault's power approach (Foucault, 1982; Giddens, 1984), the highly abstract nature of these theories creates substantial barriers to their empirical application (e.g. Moos & Dear, 1986; Sewell Jr., 1992). Giddens' structuration theory situates the analysis of power relations within an informative way of structure-agent relationship discussion and explicitly links power interactions to the broader institutional context (Giddens, 1979a, 1984). However, Giddens' conceptualization of power is limited in its ability to capture the dynamics of actors' interactions in a localized interacting logic (Geoffrey & Hodgson, 2006; Gregson, 1989). Viewing power configurations as omnipresent in social interaction, Foucault conceptualizes power as a deeply embedded, bottom-up mechanism which is nested in routine practices (Bevir, 1999; Foucault, 1998). Together, these two theoretical approaches make important contributions to the conceptualization of power, power resources and power relations. Nevertheless, in the absence of concrete and operational analytical tools to examine institutionalized processes, their empirical applications often remain confined to descriptive or narrative accounts of governance processes.

Third, existing institutionalist analytical frameworks tend either to overemphasize structural constraints or to focus excessively on actors' processes of production and reproduction. Institutionalist approaches mainly analyse how the policies serve to define, order and regulate the individuals, and how they influence the deployment of power among the actors. Structuration theory provides a significant analytical instrument for institutional analysis—namely, rules (Giddens, 1984); however, its conceptual ambiguity diminishes its utility for empirical studies (Geoffrey & Hodgson, 2006). Healey's conceptualization of urban governance improves our understanding of how structured actors, such as governments and other social actors, get institutionalized within urban development processes (Healey et al., 2002). Nevertheless, its largely unidirectional analytical logic, together with limited attention to how social practices react to and reshape institution, constrains its practical utility. Ostrom's Institutional Analysis and Development Framework delivers fruitful instruments for investigating the routinized and structured interactions within institutional settings (Ostrom, 2005). Yet its limited engagement with the duality of structure and agency, as well as with power dynamics, constrains its utility. Consequently, existing theories fail to provide a satisfactory analytical framework for scrutinizing power configurations by jointly considering reciprocal functions of broader external institutional context and localized agents' interactions.

In addition, the literature reveals four major empirical gaps. First, existing studies have not adequately conceptualized recent participatory policy reforms in China as a profound institutional transformation that fundamentally reshapes redevelopment mechanisms. These transformations include the institutionalization of property rights, constraints on rural land transfer and evolving interest-sharing mechanisms. Second, although Shenzhen was the first city in China to establish participatory redevelopment policies, empirical research on power relations in participatory redevelopment in Shenzhen remains limited. Third, current research has predominantly focused on urban village redevelopment, while relatively little attention has been paid to residential redevelopment contexts in which property rights are clearly and legally defined but fragmented among individual households. Finally, most existing studies rely on a single-case study design to examine power relations. In the absence of a systematic comparative investigation across the three major modes of participatory redevelopment—government-led, developer-led and owner-led—these studies are unable to demonstrate the overall impacts of participatory redevelopment policies on different stakeholders and the resulting configurations of power.

1.2 China's Urban Redevelopment: From Socialist State-led to Community Participatory Redevelopment

This section, based on the research questions, provides a comprehensive review of empirical studies on power relations in urban redevelopment in China. Institutional reforms have brought about critical changes in urban redevelopment mechanisms which consequently reshaped the stakeholders, their interactive behaviours, power dynamics in the development processes and project outcomes. In China, three major periods of urban redevelopment can be identified: the socialist reform period of the 1950s, the market-oriented reform period of the 1980s, and the participatory reform period since the late 2000s. Correspondingly, urban redevelopment has evolved from state-dominated redevelopment in the socialist period, the growth-coalition-dominated redevelopment during market-oriented reforms, and further to the community participatory redevelopment in the participatory period. Through a discussion of three redevelopment periods, this section identifies their distinctive characteristics from three perspectives: patterns of power relations, exogenous influences and endogenous influences. These perspectives are essential for the development of the analytical framework for identifying empirical gaps in the existing literature.

1.2.1 Socialist Reform Period—Concentrated State Power and Central Government-Dominated Redevelopment

I. Power Relations.

The socialist reform period, characterized by highly concentrated state power, spanned from the 1950s to the 1980s. Following the founding of the People's Republic of China in 1949, the Central Government initiated a series of reforms (such as the establishment of the hukou system to control free migration and the transformation of private businesses into collectively or state-owned enterprises) and quickly moved towards a centrally planned governing structure (Han, 1994). During this period, the state exercised extensive concentrated power over the planning and social-economic affairs, including urban redevelopment processes.

Empirical studies have revealed the power relations among the key stakeholders during socialist-era redevelopment. Tang (1994) demonstrated that in local urban development projects, the central state usually delegated its decisional power to relevant ministries for site selection. Upon completion of this process, state-owned and ministry-affiliated factories—designated as the planned land users—negotiated with existing occupants, such as farmers or residents, over land assembly and project construction. In this regard, designated land users implemented decisions made by state authorities and managed the land development process, while existing land users such as farmers and residents had to follow the acquisition decisions with minimal bargaining power. Other local actors, such as municipal governments and housebuilders were largely marginalized in the decision-making process. Yeh and Wu (1996) further argued that project-specific construction often led to coordination challenges. To address these issues, the central state empowered local governments to coordinate infrastructure and facilities across projects involving different ministries and departments, particularly in the later stages of socialist land development. In this process, local governments tightly controlled decision-making: they collected land requirements from work units, formulated comprehensive plans to accommodate these demands, acquired land from existing users and delegated state-owned contractors to carry out construction. Affected residents and farmers, however, remained largely powerless to influence outcomes.

Overall, existing research indicates that during the socialist period, the state—whether central or local—dominated urban redevelopment. The central state oversaw planning and resource allocation for local projects, while state-owned factories acted as local agents to execute land assembly and construction. Local governments participated primarily in coordinating project implementation at later stages. Throughout the process, affected residents and farmers held very limited power.

II. Exogenous Influences.

Exogenous influences refer to institutional and political forces that significantly shape local power structures but remain largely unaffected by individual redevelopment projects. In the context of state-dominated redevelopment, three main exogenous factors influenced local power relations. First, nationalized land ownership designated the state as the legal landowner and conferred decision-making authority over urban redevelopment projects to the central government (Tang, 1994). Second, the public land allocation system excluded market mechanisms from determining land use, effectively removing the market sector from the local power structure (Tang, 1994; Yeh & Wu, 1996). Third, centrally planned land control mechanisms reinforced the overarching authority of the central state and its sectoral agencies, thereby constraining the power of local actors, including existing land users and municipal governments (Tang, 1994).

III. Endogenous Influences.

Existing studies on state-dominated redevelopment have elucidated the sources of power for each stakeholder and the strategies they employed to exercise that power during the socialist period. The central government derived its authority from the socialist political system and reinforced its control over national projects through specific policies, regulations and development plans (Tang, 1994; Yeh & Wu, 1996). By decentralizing decision-making authority to local agencies, the central government extended its influence over land development at the local level (Yeh & Wu, 1996). Local agencies, in turn, obtained their power from the central ministries and strengthened it by forming alliances with municipal governments (Yeh & Wu, 1996). Within the centrally controlled political system, local governments primarily drew their authority from delegation by the central state, supplemented by their local social networks, administrative knowledge and jurisdictional powers (Yeh & Wu, 1996).

1.2.2 Marketed-Oriented Reform Period—Decentralized State Power and Growth-Coalition-Dominated Redevelopment

I. Power Relations.

The market-oriented redevelopment period, characterized by the decentralization of state power and the rising influence of market actors, spanned from the 1980s to the 2000s. To revitalize a society constrained by the centrally planned governance structure, the central government introduced a series of policies in the 1980s to initiate market-oriented urban reforms. These policies increased local governments' autonomy, enabling them to manage urban development processes, including the assembly of financial, land and construction resources for property-led urban projects (Hayter & Han, 1998; Yeh & Wu, 1996). Simultaneously, the reforms removed supply-side constraints on private actors, empowering them to play a substantive role in urban development (Wu, 1996; Yeh & Wu, 1996). Lacking sufficient resources to dominate decision-making independently, local governments formed growth coalitions with resource-rich actors, such as state-owned enterprises (SOEs) and government-owned or private developers, to govern urban redevelopment (He & Wu, 2005). Each powerful actor came to power in different redevelopment practices, forming three patterns in the power relations—government-led, SOEs-led and developer-led, which characterized the second redevelopment period.

Some studies argue that following market-oriented reforms, local governments increasingly assumed a dominant role in urban redevelopment. For example, Zhang (2002) observed that in Shanghai, district governments established pro-growth coalitions with non-public sectors, leading the redevelopment process while non-public actors leveraged their growing economic power to influence governmental decisions.

Residents, however, were largely excluded due to limited electoral influence and the absence of representative institutions. Similarly, Han and Wang (2003) found that in Chongqing, local governments maintained a dominant role in land redevelopment, while developers acted as initiators who had to comply with government directives to secure support. Shin (2009) noted that after the introduction of affordable housing policies in Beijing in 2000, local governments proactively dominated residential redevelopment, providing project funding and transforming affiliated agencies into development corporations with comprehensive authority over planning, land assembly, construction and relocation. Communities, in turn, wielded minimal power and were largely unable to influence decision-making.

A handful of studies on the topic pointed out that SOEs, as the land owner, were able to deploy strong power to influence, even change the local government's decisions in some cases, although local governments were configured with notable power in decision-making. For example, Zhu (1999) showed that in Shenzhen, municipal governments granted some SOE construction companies full developer status, enabling them to manage land transfers and make development decisions. With limited financial resources, local governments relied on land subsidies to leverage market actors. In his succeeding research, Zhu (2002) revealed that based on their land ownership and consequent strong power in the decision-making structure, SOEs transformed to a market-oriented role in the pursuit of economic interest through commanding their land assets in urban (re)development. Fang and Zhang (2003) argued that in Beijing, in order to avoid sharing interest with the central government, local governments deliberately allocated land to SOE developers for redevelopment at a price far below the market rate. In this process, the municipalities were able to seek massive monetary profit and even public housing from the residential redevelopment project. Based on their land resources and a close relationship with the local government, SOEs were powerful to lobby the governments to change land use and overturn zoning and building codes to enlarge the profit-seeking space for developers. The community had no power to influence the decision-making process without ownership of their housing unit. Han and Wang (2003) mentioned in Chongqing, SOEs, occupying a large proportion of urban land, played the role of land provider in some urban redevelopment projects. Being short of capital but rich in administratively allocated land, they either built up their own development companies or cooperated with private developers to accomplish and enormously profit from the property development.

Facing increasing constraints of financial resources in the rapid urbanization process, the local government had to facilitate the deployment of market sectors (Wu, 2004). Several studies demonstrated that the mounting power of the private developers, especially the international real estate companies, outweighed the local government in the redevelopment process, modifying the involved participants and the governance structures. For example, He and Wu (2005) described that in a landmark redevelopment case in Shanghai, the international developer came to power to take over the redevelopment process from investment and design to construction and management, based on its ample capital, expertise and development experiences. The local government played the role of facilitator to set up the redevelopment project as a hallmark with political objectives and mobilized massive administrative resources to prioritize and smooth the way for the project. The residents, with no arguing power, had to follow the decision of the economically powerful actors. Yang and Chang (2007) indicated that in Shanghai, a joint venture between the international developer and government-owned development company was established to conduct the redevelopment project. The foreign developer was configured with strong power to influence the decision-making process for its vast majority of investment share in the joint venture. With a small portion of investment through a state-owned real estate company, the local government was able to participate in the entire decision-making process and played a co-operator role. The finding in Shin (2009)s research agreed with Yang and Chang (2007)'s that in a redevelopment project in the central area of Beijing, overseas investors took charge of the urban redevelopment by forming a joint venture with locally based developer. The local government played a facilitating role by not only building up an alliance between the foreign and local developers but also arranging the land lease through negotiation rather than public auction.

Subordinated stakeholders, including state banks and professional demolition companies, were identified in current studies. For example, Wu (2004) argued that real estate companies relied heavily on loans from state banks to initiate the redevelopment project. Han and Wang (2003) identified the loan of state banks significantly affected the developer and the construction firms' investment decisions. Also, Wu (2004) indicated that real estate companies assigned the work to specialized demolition companies to force residents to move out, as a result of the complexity and difficulty of demolition work.

The above literature presented the power relations among the main stakeholders nested in the urban redevelopment process after the market reform in China. The marketization of land and decentralization of power from the central government have positioned local governments onto the vital role in urban development with essential influences on the decision-making process (He & Wu, 2009). With the mounting pressure of economic growth, the local government gradually turned into an entrepreneurial role to create an investment environment and minimized its responsibility of maintaining social stability. To profit from local redevelopment projects, the local government, on one hand, formed alliances with the SOEs to evade the ban of its involvement in local development projects (Fang & Zhang, 2003; Han & Wang, 2003); on the other hand, it resorted to market power to accomplish the redevelopment when facing absence of financial and technical resources (He & Wu, 2005; Yang & Chang, 2007). In this process, local developers (either government-owned or private) often assumed a moderate role to implement the powerful actors' decisions (Fang & Zhang, 2003). International developers, configured with ample expertise and capital resources, came to power to dominate the redevelopment in some landmark projects (He & Wu, 2005; Shin, 2009; Yang & Chang, 2007). The leading role in power structure thus was alternated among the three economically powerful actors including the local government, the SOEs and the developer, depending on the practical and local context of the redevelopment projects. These powerful actors usually formed coalitions to necessitate their substantial control in prompting urban growth, using urban development as one platform for profit extraction. Whereas, the community, which was configured without any evident power, played a marginalized role (Shin, 2009; Yang & Chang, 2007).

II. Exogenous Influences.

A number of studies highlighted that through an array of policies, the central government transformed the planned economy to a market-oriented one. These changes had crucial impacts on the localized power relations in urban redevelopment. For example, Dowall (1994) indicated that the recognition of property ownership in housing reform empowered the residents to negotiate with the local government. Both Fang and Zhang (2003) and Yang and Chang (2007) revealed that the progress of the market reforms strengthened the local governments' control over urban redevelopment and invited market sectors to play a role in the decision-making process. Hsing (2010) and Yang and Chang (2007) pointed out that the prohibitive policy of land use transfer between existing and potential land users enabled the local states to monopolize the primary land market, which significantly empowered the local government and compromised the power of SOEs. However, treating the institutional reform as a singular force, these studies failed to depict the overall institutional changes and their impacts on the localized power relations from a perspective of the whole picture. To address this gap, He and Wu (2009) stated that the decentralization and the rescaling of the state associated with land and housing reforms shifted the urban redevelopment from an approach of welfare provision to a way of growth promotion, reconstituting the redevelopment mechanisms and the involved stakeholders.

Two studies revealed the influences of local land use and planning policies on the local power configuration. One is by Dowall (1994), who considered the local redevelopment policies emphasizing in-kind compensation in the 1990s empowered the affected residents to obtain satisfying compensation. The other is by Shin (2009), who indicated that the policy promoting the supply of affordable housing in Beijing allowed the local government and the affected residents to finance the project jointly. This policy enabled residents to join in the pro-growth coalition and empowered them to obtain satisfying compensation.

Both physical and social attributes of the community have been identified in current studies. Communities with premium location and large size, as physical attributes, granted certain power to the affected residents when negotiating with the local government or developers (Yang & Chang, 2007). Hsing (2010) found that long-time neighbours, as a social attribute, could enhance residents' collective actions towards the strategies of the local government and the developer in residential redevelopment.

The above studies revealed that exogenous influences that had impacts on local power relations including the following references: central policies towards institutional changes, local policies in land use and planning field and physical attributes (premium location) and social attributes (long-term neighbours) of the community.

III. Endogenous Influences.

Current literature revealed that local governments were able to obtain their power from various resources in the growth-coalition-dominated period. After the marketed-oriented reform in the late 1980s, local governments received decentralized decision-making power from the central state; it thus became the formal land owner with access to land resources (Fang & Zhang, 2003; Zhang, 2002). The land ownership provided the local state with the regulatory authority over redevelopment planning to approve, suspend or cancel the project (Han, 2000; Shin, 2009; Yeh & Wu, 1996). The land use and planning system granted the local government the administrative jurisdictions on land use and development intensity (Dowall, 1994; Han & Wang, 2003). The local government also got its power from the control of financial institutions (Han & Wang, 2003) and a small portion of investment in redevelopment projects (Yang & Chang, 2007).

In the local interaction process, it is recognized that the local government adopted strategies to maintain and strengthen its power in urban redevelopment. The municipalities established government-owned development companies or other public agencies as a state-endorsed representative with partial administrative and approval power (Han, 2000; Yeh & Wu, 1996; Zhang, 2002). The provisional agency took charge of mobilizing political forces to expel residents, under the financial and administrative support of the local government (Fang & Zhang, 2003). Local governments also placed a premium on some landmark cases so that it could pour its massive administrative resources to facilitate the project (He & Wu, 2005; Shin, 2009; Yang & Chang, 2007). When facing the absence of financial capability or expertise, the local government attempted to form a coalition (such as joint venture companies) with other economically powerful actors, such as SOEs and international developers to coordinate more resources for project facilitation (e.g. Yang & Chang, 2007; Zhu, 1999). Within the coalition, the local government played a facilitating role in creating a business climate and deployed its power through land subsidies (Zhu, 1999).

As the legal land owner, SOEs mainly derived their power from the decision-making right about the transfer of the land they occupied (Han & Wang, 2003; Zhu, 1999), autonomous authorities about planning (Zhu, 2002) and good relationships with local governments (Fang & Zhang, 2003). They initiated and promoted the redevelopment through either establishing self-owned development company (Zhu, 1999, 2002), or cooperating with private developers and the local authority (Fang & Zhang, 2003).

Three kinds of developers were identified in the growth-coalition-dominated redevelopment process. Each derived power from different resources and resorted to different strategies and actions to maintain and strengthen their power. For the government-owned developer, its power exclusively stemmed from the decentralized land development right (Han, 2000; Zhang, 2002). Local private sectors obtained its power from their financial capital, local knowledge and a good relationship with the local governments (Han & Wang, 2003). They cooperated with financial institutions to strengthen their capital resources and power (Han & Wang, 2003). International developers garnered their power from ample expertise such as advanced design skills (Yang & Chang, 2007), sufficient financial resources (He & Wu, 2005; Shin, 2009), good reputation (Yang & Chang, 2007) and even under-the-table deals with the local government (He & Wu, 2005; Shin, 2009). In practice, these developers all tended to maintain a good relationship with the local government and collaborate with professional demolition companies to accelerate the project and reduce the cost (Wu, 2004).

Residents' power was influenced by several factors, legal housing ownership, working in local government and other public institutes and involved in hallmark projects. Legal housing ownership significantly empowered the affected residents in urban redevelopment (Fang & Zhang, 2003; Yang & Chang, 2007; Zhang, 2002). Shin (2009) found that in Beijing, residents who worked in the local government and other public work units could obtain better compensation on relocation. Besides, residents were temporarily better-off in their compensation when the local government started the project for political objectives (He & Wu, 2005).

Residents also took various strategies and actions to deploy their power in urban redevelopment. They tended to find space for mobilization within policy gaps and resorted diverse strategies according to their interpretation and anticipation of state action, such as individual or collective lawsuits, petitions, violent confrontations and cross-neighbourhood mobilization (Hsing, 2010). Sometimes, the individual and mobilized resistances they adopted were approved to be futile when facing the powerful coalition between developers and local governments because no public agent was representing their interests (Shin, 2009; Zhang, 2002).

1.2.3 Participatory Reform Period—Rising Community Power and Participatory Urban Redevelopment

I. Power Relations

Current studies suggest that a participatory redevelopment has been formed recently featured by the rising community power. A recent policy *San Jiu Gai Zao* is promulgated by the central government to move away from expansion in urban fringe areas towards the intensification of built-up urban areas in Guangdong province. The new policy led to a power restructuring process in urban redevelopment where the power of the economically powerful actors, such as local government and the developer, is constrained and the power of the residents is strengthened (Li & Liu, 2018; Tian & Yao, 2018). With increased power to argue against the economically powerful stakeholders, residents can play an increasingly important role in the decision-making process (e.g. Chung, 2013). Increasingly inclusive and participatory governance is emerging under the new power structure (Zhang et al., 2016). The traditionally powerful actors such as the local government and the developers as well as the newly empowered property owners thus pose their influences in various ways, leading to three patterns of power relations in urban redevelopment, namely, the government-led, developer-led and owner-led redevelopment.

A handful of studies revealed that in some redevelopment cases, the local government played a dominant role while residents were configured with ascending power to counteract the influences of the local government. Cheng (2012) revealed that in Fangzhicheng, Xi'an, by forming a coalition with a developer, the local government intended to undermine the interests of the affected residents and compellingly implement the redevelopment. Residents, despite without formal property deeds, formed mobilized actions to resist the repression of the coalition between local authorities and developers. As the forceful measures failed to achieve the land assembly, the local government had to adopt a collaborative attitude to disintegrate the collective resistance and retain supports from the affected residents. On completion of agreements from the majority of the affected residents, it thus was able to initiate the demolition steadily. Zhai and Ng (2013) contended that in a historic Muslim community in Xi'an, after proposing a redevelopment plan, the local government strived to create a market environment to attract private investment for its prompt implementation. Affected residents organized mobilized actions to resist the local government while representatives repeatedly submitted collective petitions to higher-level governments for political assistance. As a response, the Ministry of Construction delegated investigators to inspect the legitimacy of the redevelopment plan at local level. The intervention of the central government successfully prevented the implementation of the demolition plan. Zhang et al. (2016) found that to lift the land value of the central city in Beijing, the local government formed a coalition with a SOE, the de facto landowner, to redevelop the Jiuxianqiao area. Residents established a self-organized committee to mobilize collective resistance to argue against the alliance, generating social instability. To facilitate the Olympic Games in Beijing, the central government pressurized the local government to prioritize social stability. The latter thus organized a public vote to collect residents' opinion and suspended the redevelopment based on the result of the public vote. Two years later, the local government restarted the redevelopment and allowed residents to participate in the decision-making process. An inclusive governance was thus formed to facilitate the redevelopment implementation.

Studies indicated that although residents were significantly empowered to argue with the economically powerful party, with the cooperation of the local government, the developer sometimes played a leading role in participatory redevelopment. For example, Li and Li (2011) found that in the redevelopment of Yunong Village, Shenzhen, after realizing the forceful implementation was no longer practicable, the local government waived its overwhelming power and turned to adopt a collaborative approach to implementing the redevelopment. It facilitated a tri-partner collaboration among the local state, the village collective and a reputable developer. The private developer was delegated to play a more active role to negotiate with the affected villagers on the redevelopment plan and compensation details. Villagers, as the de facto landowner, obtained premium compensation in the process. To protect the interests of the developer, the local government permitted an extraordinarily high development density to increase marketable apartments for the former. Cheng (2012) found that in Foshan, the local government excluded other competitors in the land auction to protect the involvement of a reputable international developer. Once obtaining the land redevelopment right, the developer was authorized to play a critical role in the redevelopment process, taking control of the procedures from establishing redevelopment plan and design to construction and sale. The local government played a facilitating role, on one hand, to revise the redevelopment plan according to the developer's requirement; on the other hand, it pressurized the affected residents to relocate in order to prepare the land for redevelopment. The affected residents, holding the legal ownership, had stronger arguing power to negotiate with the government to obtain satisfied compensation standards.

Several studies indicated that the community was authorized with self-redevelopment rights to initiate the owner-led redevelopment. Cheng (2012) contended that in the redevelopment of Wenchong village, Guangzhou, the local government became less dominant and transformed to be a facilitator. To smooth the way of urban village redevelopment, it delegated full administrative and approval power to the village collective company and made the latter as a critical role to negotiate with the developer. The village collective company thus turned to play a dual role in the process. On one hand, it negotiated with the developer to raise compensation terms for the village collective and individual villagers; on the other hand, it strived to convince all the villagers to sign the agreement to facilitate the redevelopment. In this process, villagers' requests were adequately responded, such as raising compensation standards and replacing the previous developer with a larger one with greater capacity and reputation. Chung (2013) indicated that the Guangzhou government played the role of regulator and adopted a soft approach to promoting urban village redevelopment. In Xinxiang village, native villagers formed a Village Joint Venture Enterprise (VJVE) as a collective representative of the village to negotiate with the local government and developer in the redevelopment process. As the legal land owner, the VJVE, on one hand, played a power broker of the local government to fight for interests for both individual villagers and the collective. On the other hand, it undertook an entrepreneurial role by not only forming a partnership with the private developer for extracting profit but striving to persuade individual villagers to follow its growth-oriented redevelopment plan. In this process, the local government deployed its administrative power to reconciliate the confrontation between the village collective and uncooperative residents to facilitate the redevelopment, while the developer utilized its capitalist power to seek profit from the redevelopment process. Zhou (2014) revealed that in Liede village, Guangzhou, an urban village close to the new CBD area, the local government established a government-led redevelopment plan to obtain the village land to prepare the 2010 Asian Game. Under the inspiration of the *San Jiu Gai Zao* policy, the VJVE refused government's redevelopment scheme and proposed a self-redevelopment agenda. The local government approved the self-redevelopment agenda and began to play a facilitating role. The

VJVE formed a collaborative partnership with the local authority and developers to complete the redevelopment.

The above studies contended that the affected residents, a traditional powerless and excluded role, begin to play an essential role in the decision-making process in urban redevelopment, forming a new era of participatory redevelopment. Power relations among the main stakeholders have converted to be more inclusive and collaborative in urban redevelopment. With increasingly collaborative institutional transformation, the local government's power was constrained by the mounting community power (Zhai & Ng, 2013; Zhang et al., 2016). In urban village redevelopment projects, the local government became less dominant and powerful. The government, on one hand, played a regulator to establish redevelopment policies and supervise the implementation process (Chung, 2013); on the other hand, it worked as a facilitator to smooth the way of land transfer and redevelopment implementation by forming partnership among itself, the collective villages and developers (Cheng, 2012; Chung, 2013). The villagers, as de facto land owners, were authorized significant power to influence the decision-making process (Chung, 2013). VJVEs were formed by villagers as representatives to negotiate with the traditional economically powerful actors, the local government and developers, to pursue profits (Zhou, 2014). When confronting conflicts between collective and individual interests, the VJVEs turned to an entrepreneurial role to convince the uncooperative villagers to sign contracts (Cheng, 2012). Facing the pressure from the VJVEs, the individual villagers had ample arguing power to decide whether to accept the compensation offer or not, which had a significant impact on the project (Cheng, 2012). The developer played a role of initiator to form a negotiation platform with the VJVEs. Once both parties reached a consensus of the redevelopment schemes, the developer assumed a role of builder to implement the reconstruction.

However, significant and evident knowledge gaps in power relations remain in current studies. Existing research on government-led residential redevelopment failed to reveal the developer's role and its power configuration to react to the mounting community power. Meanwhile, as participatory urban redevelopment policies constrain the government interventions in urban redevelopment and enable the direct land transfer between developers and existing land users, developers, urban villagers and urban residents are empowered to initiate redevelopment projects (Li & Liu, 2018; Tian & Yao, 2018). However, current studies on power relations in terms of developer-led and owner-led redevelopment mainly focus on urban village redevelopment projects in which the village collective possessed land ownership and thus had strong power to argue with the local government (Chung, 2013). Few studies have been able to draw on any empirical research into the power deployment in developer-led and owner-led residential redevelopment, in which individual household fragmentedly owned the properties.

II. Exogenous Influences

Three exogenous influences that have important impacts on power relations among the primary stakeholders in urban redevelopment have been identified in current studies: central reform policies, local land use and planning policies and physical community attributes. Firstly, a handful of studies revealed the influences of the *San Jiu Gai Zao* policy on the local power configuration in urban redevelopment. For example, Ye (2011) indicated that in 2009, the Guangdong province issued its *San Jiu Gai Zao* policy to facilitate urban redevelopment practice. This policy turned to constrain the power of the local government and made it a regulator and facilitator in the decision-making process. Meanwhile, it empowered the affected residents (property owners) to join in the decision-making process to obtain fair compensation and relocation. Lin (2015) indicated that the *San Jiu Gai Zao* policy removed the constraints of the developer from the supply side because it allowed the direct land transfer between existing users and developers through negotiation. As a consequence, the existing property owner can directly participate in the decision-making process. The developer can play a more flexible role in urban redevelopment to sufficiently motivate the market power. Li and Liu (2018) found that the *San Jiu Gai Zao* policy is able to promote alliance formation among the involved stakeholders (the state, the market and communities) to facilitate urban redevelopment. Tian and Yao (2018) revealed that to motivate existing land users and pacify social resistance in urban redevelopment, the *San Jiu Gai Zao* policy granted self-redevelopment rights to existing land users. Schoon (2014) regarded the *San Jiu Gai Zao* policy as a core reform in land development in China since the opening-up and market reform in 1978. It changed the objectives, involved stakeholders and mechanisms of urban (re)development. However, current studies mainly regard that the participatory redevelopment reform was triggered by a singular force—*San Jiu Gai Zao* policy—that changed redevelopment mechanisms and reconstituted power relations among stakeholders in urban redevelopment. None of them investigated the recent transformation from a standpoint of a profound institutional reform associated with significant changes in multiple aspects, including constraint of rural land transfer, institutionalization of property right, new land transfer approach and fairer implementation procedure and interest distribution.

There is one study indicating the impact of local land use and planning policies on power relations in urban redevelopment. Zhou (2014) noted that in Guangzhou, the abolishment of the prohibitive regulation allowed developers to participate in urban village redevelopment in the late 2000s. The policy in 2012, however, prevented developers' direct participation from the urban village redevelopment in Guangzhou, substantially diminishing developer's power. Given that increasing local land use and planning policies emerge in participatory period, studies on their impacts on power relations are rather limited. What the local land use and planning policies are and how they restructure power relations in participatory redevelopment need further examination.

Current studies demonstrate the physical attribute of the community has an important impact on local power relations. Chung (2013) contended that the premium location and large scale of village empowered the collective company to obtain better compensation in the negotiation with the developer. Zhou (2014) indicated that the Liede Village's adjacency to CBD area and the pragmatic demand of the local government to prepare for the 2010 Asian Game motivated the municipality to approve the self-redevelopment application from the affected residents. In addition to physical attribute, existing literature fails to pay attention to the social attribute of the community. More research on the influence of social attribute of the community is necessary.

III. Endogenous Influences

Current studies on the empowerment process of the different stakeholders in participatory redevelopment mainly focus on the urban village redevelopment. In this process, the local government primarily obtained its power from its ruling position (Chung, 2013) and jurisdictions on land use and development intensity (Lin, 2015). The local government also executed some strategies and actions to strengthen their power to influence the decision-making process. For example, the local government, which is no longer the legal land owner, formed a partnership with the VJVEs and executed its influences through a team of staff working closely with the latter (Li & Li, 2011). It also deployed its power to harmonize the relationship among the village collective, individual villagers and developers towards project implementation (Chung, 2013; Zhang et al., 2013).

For the indigenous villagers in the urban village, they derived their power from de facto land ownership (Chung, 2013), institutional authorization of self-redevelopment right (Chung, 2013; Zhou, 2014), local identity and place attachment (Chung, 2013) and existing social networks (Chung, 2013). To strengthen their power, villagers formed a collective-owned company as the legal land owner to representatively negotiate with local authorities and developers, bargaining for fair treatment and compensation (Li & Li, 2011; Zhou, 2014). When the individual interests contradicted to the collective interests, some villagers resorted to mobilized actions to resist the village collective's decision, such as protests, demonstrations and journalists invitation to strengthen their power (Cheng, 2012; Chung, 2013).

Developers in the redevelopment of urban villages derived their power from ample capital resources and expertise (Li & Li, 2011), large company scale and reputation (Cheng, 2012), local relations (Li & Li, 2011), successful redevelopment experience (Li & Li, 2011) and the most important, policy-defined participatory rights (Chung, 2013). For example, Cheng (2012) found that developers in small scales were equivalent to ones with limited capabilities and thus incurred distrust from the affected villagers while national developers with large scales and good reputation would be regarded as capable and with ample expertise. Chung (2013) argued that the removing of restriction on the developer's participation significantly empowered the developer to execute their capitalist resources in urban village redevelopment. Developers, in this process, tended to form a coalition with the VJVEs to secure their profits from the projects.

With limited research on the participatory residential redevelopment, significant and evident gaps remain. For local government, current studies indicate that they established a state-endorsed office on site with almost full administrative and approval powers to negotiate with the affected residents (Cheng, 2012; Zhai & Ng, 2013). The public agency adopted strategies to dissolve the power of the residents, including media censorship, monetary incentives and legal actions (Cheng, 2012). However, where does the power of the local government come from? What strategies and actions did the local government use to maintain and strengthen its power in urban redevelopment? These questions need to be answered.

For urban residents, studies demonstrate that they executed various strategies and actions to strengthen their power in participatory redevelopment. Cheng (2012) indicated that the residents resisted the economically powerful actors through lodging individual and collective actions, initiating protests, posting online petitions, exposing their experience to journalists and self-training of relevant policies. Zhang et al. (2016) found that to pressurize the local government, residents established a self-organized committee to mobilize collective resistance and petitioned to the central government for direct central intervention in Beijing. Zhai and Ng (2013) contended that in Xi'an, affected residents organized not only mobilized actions to resist the local government but also submitted collective petitions to higher-level government for political assistance.

Current research fails to demonstrate how urban residents obtain their power to make a difference in residential redevelopment. Different from the original villagers, the urban residents were configured with the property ownership rather than the land ownership; and the social networks and local identity between the urban community and the urban village are also highly different. Empirical research on the power dynamics of urban residents is in urgent demand.

Knowledge gaps in the role of developers and their empowerment condition in residential redevelopment are evident. Cheng (2012) found that the ample financial capability of the developer enabled them to form a coalition with local governments, which empowered them to obtain priority in accessing land and redevelopment projects. He also argued that the international developer's successful and remarkable cases in first-tier cities (e.g. Shanghai) substantially empowered it to negotiate with the local governments in second-tier cities (e.g. Foshan). Zhai and Ng (2013) indicated that in Xi'an, the emerging urban land market empowered private developers to join in redevelopment projects. These developers were also motivated to participate by the local government's endeavours to create an investment environment. Zhang et al. (2016) contended that facing unpredictable risks of urban communities with ambiguous property ownership condition, the local government delegated the SOEs, the legal landowner, to initiate the redevelopment. As inexperienced in redevelopment projects, the latter resorted to private developers with ample capital and expertise to accomplish the redevelopment. The above studies revealed that developers obtained their power from their financial resources, expertise, reputation and successful redevelopment experience. As the participatory redevelopment policy enabled direct land transfer between developers and residents, more research is needed to examine the influences of participatory policies on the role of private developers. Current studies mainly revealed the empowerment of developers in the government-led redevelopment, where the local authorities took the lead to negotiate and persuade the affected residents to be cooperative. Not clear are the strategies and actions the developer adopted to react, even counteract the rising power of the community in the developer-led and owner-led redevelopment process.

1.2.4 Summary

Table 1.1 summarizes Power Relations of Urban Redevelopment across the Three Stages. While a growing body of literature documents the transformation of power relations in participatory urban redevelopment—such as the declining dominance of local governments and the increasingly active involvement of affected residents and villagers—the role of private developers, as well as their changing weight in negotiation processes, remains insufficiently understood, particularly in government-led redevelopment initiatives. Existing studies suggest that participatory policies have empowered both developers and urban residents to initiate redevelopment projects; however, few have systematically examined power relations in developer-led and owner-led redevelopment within communities where property rights are clearly defined on state-owned land.

Most current studies tend to attribute the transformation of power relations to a single policy driver, namely *San Jiu Gai Zao*. In doing so, they overlook the broader and more profound nature of institutional reform that has reshaped redevelopment mechanisms from multiple perspectives, including the institutionalization of property rights, constraints on rural land transfer and evolving interest-sharing arrangements. Moreover, compared with the extensive discussion of central-level policies, local land use and planning regulations—and their concrete implications for power relations—have received limited scholarly attention. Existing research also lacks a systematic framework for categorizing exogenous and endogenous influences, as these dimensions are often examined in a fragmented and unsystematic manner.

Empirically, studies on power relations in participatory redevelopment have predominantly focused on cities such as Guangzhou, Beijing and Xi'an. Shenzhen, despite being a well-recognized policy experimentation ground and having introduced participatory redevelopment policies as early as 2007, remains underexplored in this regard. Empirical investigations into the dynamics of power relations in Shenzhen's participatory redevelopment are still scarce.

In addition, most existing research relies on single-case study designs, which limits the ability to capture the systemic and comprehensive impacts of participatory redevelopment policies on stakeholder interactions and power configurations. Although Cheng (2012) adopted a multi-case approach by examining three redevelopment projects, his study fell short of revealing the dynamic evolution of power relations in urban redevelopment for two main reasons. First, the three cases were located in different cities with distinct political contexts and physical conditions, varying significantly in scale and type—ranging from a large SOE-owned old community to a medium-sized dilapidated residential area and a large urban village. This heterogeneity undermined the comparability of the cases and constrained the analysis of localized power relations. Second, by failing to examine how exogenous and endogenous influences shape local power relations, the study remained largely descriptive, focusing on redevelopment processes rather than uncovering how institutional

Table 1.1 Power relations of urban redevelopment across three stages

Redevelopment type	State-dominated redevelopment	Growth-coalition-dominated redevelopment	Participatory redevelopment
Period	The 1950s to 1980s	The 1980s to 2000s	The 2000s to present
Key actors	Central state Ministry-owned factories Municipality Developers Tenants	Local government SOEs Local developers (private or government-owned developer) International developers Residents	Local government Developers SOEs Village collective (generally in terms of VJVE) Indigenous villagers The affected residents
Power relations	State-dominated	Government-led SOEs-led Developer-led	Government-led Developer-led Owner-led
Exogenous influences	Nationalized land ownership Public land allocation system Central-planned land control mechanism	Central policies towards institutional reform (Housing reform, land reform, decentralization and rescaling of the state and prohibitive policy of land use transfer) Local land use and planning policies (redevelopment policy emphasizing in-kind compensation, affordable housing policy) Physical attribute (premium location and large size of the community, long-time neighbours)	*San Jiu Gai Zao* policy Local land use and planning policies (regulations about the involvement of developers) Physical attributes (premium location)

(continued)

Table 1.1 (continued)

Redevelopment type	State-dominated redevelopment	Growth-coalition-dominated redevelopment	Participatory redevelopment
Endogenous influences	**Central government** Power resources: Socialist political system Strategies: Policies, regulations, development plans and employment of local agencies **Ministry-owned factories** Power resources: The delegation of the central ministries Strategies: Form alliances with local municipalities **Local government** Power resources: Commission of the central state, local social network, local knowledge, administrative jurisdictions Strategies: Coordinate construction for separated projects	**Local government** Power resources: Regulatory authority over redevelopment planning, administrative jurisdictions, land ownership, control of financial institutions, a small portion of investment in redevelopment projects Strategies: Established public agencies to promote the progress of urban redevelopment, devote administrative resources to landmark cases, form coalition with other resource-rich actors **SOEs** Power resources: Land ownership, good relationships with local governments, autonomous authorities about planning Strategies: Establish self-owned development company and cooperate with private developers and the local authority. **Developers** Power resources: Decentralized land development right (government-owned developer), financial capital, local knowledge and a good relationship with the local governments (local private developer), ample expertise, financial resources, good reputation and relations with the local government (international developer) Strategies: Maintain the good relationship with the local government, collaborate with professional demolition companies to accelerate the project **Residents** Power resources: Legal housing ownership, working in local government and other public institutes and involved in hallmark projects Strategies: Find space for mobilization within policy gaps and resorted diverse strategies according to their interpretation and anticipation of state action, such as individual or collective lawsuits, petitions, violent confrontations and cross-neighbourhood mobilization	**Local government** Power resources: Definition of regulations and policies, jurisdiction on land use and development intensity Strategies: Form a partnership with the VJVEs and extend influences through a staff team, harmonize relationship among the main participants **Indigenous villagers** Power resources: Land ownership, institutional authorization of self-redevelopment right, local identity and place attachment, existing social network Strategies: Form the collective-owned company as a representative to bargain for fair treatment and compensation, mobilized actions to fight for individual interests **Developers** Power resources: Capital resources, expertise, reputation, local relations, successful redevelopment experience, policy definition Strategies: Form coalition with VJVEs to secure their profits

changes reconfigure actors' empowerment and the resulting power dynamics. Consequently, a systematic and comparative investigation of different patterns of power relations in participatory urban redevelopment in China remains notably absent.

1.3 Objectives and Questions of Research

1.3.1 Key Terminologies

Power and Power Resources.

Power in this book refers to the transformative capacity enabling an actor to act otherwise to achieve their ends through making influences others in the decision-making process (Giddens, 1984). The deployment of power is grounded in the mobilization of power resources. Drawing on structuration theory, this research categorizes power resources into three types: structural resources, allocative resources and authoritative resources. Structural power derives from the configuration of structural properties such as institutions and policies while the agential power is constituted through the possession and mobilization of allocative and authoritative resources (Giddens, 1984). Allocative resources refer to control over material facilities such as financial strength and technical expertise. Authoritative resources consist of the human command over people, including capacities for self-development and self-expression such as knowledge, skills and experts and organizations and relations of human beings in mutual association such as belief, social norms, reputation, authority, social relationships and networks (Cohen, 1989; Giddens, 1984; Phipps, 2000).

Rules

Rules refer to channels that the power emits through the configuration of the strategies and actions. The characteristics of rules signify diverse manifestations and effects of the adopted strategies and actions to structure and restructure the power relations. The categorization of rules by Ostrom proposes an approach to precisely structuring how power functions (Ostrom, 2005). In varying forms, rules are applied to an array of actions, such as the use of (1) position rules to specify positions for participants in development process, (2) boundary rules to define how participants are chosen to or leave these positions, (3) choice rules to constrain other actors' action, (4) scope rules to allow or forbid the effect of other actor's action, (5) aggregation rules to unite actors towards a collective action, (6) information rules to define approaches of communication among actors, and (7) pay-off rules for the distribution of interests and costs among participants (Ostrom, 1986, 2005).

Power Relations

Power relations are conceptualized as a dynamic and continuously relational process which are composed of the fluid procedures and activities from strategies and actions of the actors with recurrent changes (Foucault, 1982, 1998; Giddens, 1984). Constituted through a complex configuration of rules and resources, power relations operate as a bottom-up force through which institutions are mutually constituted and transformed. This process reveals how actors react to, adapt to and reproduce structural constraints.

Power Arena

This book disaggregates power relations into a dynamic sequence of cross-sectional relational "power arenas" in order to unpack decision-making processes into distinct analytical components. This disaggregation facilitates the identification of varying configurations and accumulations of social relations that emerge within each arena. Each arena is treated as a focal analytical unit within the redevelopment process, centring on who is empowered to act or to influence outcomes within a specific decision domain.

Multiple arenas may be linked sequentially or operate simultaneously, and they are mutually dependent, collectively constituting the broader social reality of redevelopment. A power arena is formed when participants share access to a common set of information relevant to a particular decision. Within each arena, stakeholders—including local governments, developers and affected residents—interact with one another according to their respective interests and values.

Power Arena Framework

Figure 1.1 illustrates the power arena analytical framework, which integrates Giddens' conception of power embedded in the structure–agency relationship, Foucault's Power approach with Ostrom's Institutional Analysis and Development (IAD) framework on interconnected action arenas. This synthesis provides a novel analytical lens that moves beyond a static depiction of power relations by reconceptualizing them as a dynamic sequence of cross-sectional "relational arenas", within which project stakeholders interact over time.

Within each arena, interactions are shaped by both exogenous and endogenous influences, while the outcomes of power interplay are evaluated through three criteria—equity, institutional adaptability and transaction costs—corresponding respectively to redistributive effects, path dependence and process efficiency in redevelopment outcomes. These (re)development outcomes subsequently feed back into

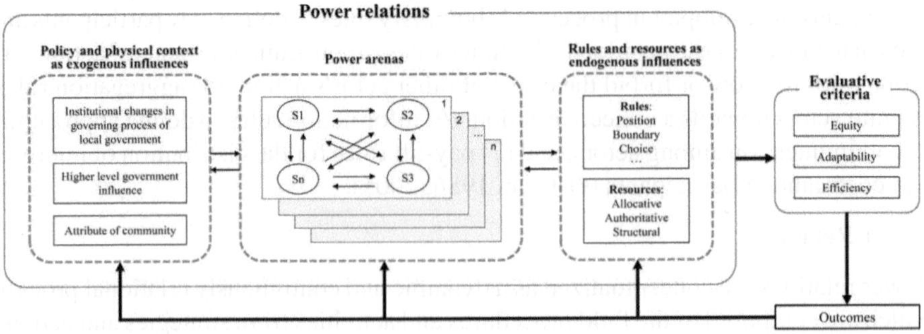

Fig. 1.1 A power arena framework

the arenas, as well as into exogenous and endogenous conditions, forming an evolving social–communicative mechanism.

Exogenous influences primarily include policy frameworks and institutional contexts at both local and central government levels, together with community attributes. Endogenous influences consist of rules and resources. Within each power arena, stakeholders (S1, S2, … Sn) interact according to their respective interests and values. Endogenous forces condition what decisions actors are authorized to make, how decisions are made, what actions are undertaken, and the consequences these actions generate. The three evaluative criteria are adapted from Ostrom (2005): equity assesses the distribution of interests among stakeholders; adaptability captures actors' capacity to respond to shifts in power deployment; and transaction costs reflect the overall performance and efficiency of the redevelopment process.

It should be noted that endogenous rules and resources are not independent of exogenous forces. Policies and institutional contexts define the action situation, while rules—operating as principles, action channels and outcome-shaping mechanisms—interact with allocative and authoritative resources to structure the unfolding of power relations. Rules in diverse forms may be applied across different stages and types of action (Ostrom, 2005). Overall, the power arena analytical framework enables a micro-level evaluation of both the distributive context and outcomes of urban redevelopment processes.

1.3.2 Research Objectives

To address the existing theoretical and empirical gaps, this book draws inspiration from Ostrom's Institutional Analysis and Development (IAD) framework, Giddens's structuration theory, and Foucault's approach to power to construct a power arena analytical framework. Within this framework, power relations are examined through the combined effects of both exogenous variables (institutional factors) and internal variables (the interaction between power resources and rules). Furthermore, power relations are disaggregated into a dynamic sequence of cross-sectional "relational arenas" in order to examine decision-making processes in a fine-grained and systematic manner.

Empirically, the study selects three cases of dilapidated residential redevelopment in Shenzhen—representing government-led, developer-led and owner-led modes—to conduct a systematic investigation of power relations across three patterns of participatory redevelopment. This comparative design enables the book to demonstrate the overall impacts of participatory redevelopment policies on involved stakeholders and the resulting configurations of power.

In this regard, the research is organized to meet three specific objectives:

1. To identify the key stakeholders and assess the ongoing power relations among them in the participatory redevelopment process;

2. To investigate both exogenous and endogenous influences that drive changes in power relations;
3. To evaluate the impacts of different configurations of power relations on the processes and outcomes of participatory redevelopment;

The first two objectives will be addressed through the investigation of three corresponding research questions, while the third objective will be achieved by systematically evaluating the various forms of power relations and their associated outcomes.

1.3.3 Research Questions

This book seeks to address the following three questions:

1. How is power distributed among the primary stakeholders in participatory redevelopment, including the local government, developers and property owners?
2. What are the exogenous influences and how do they impact the power relations?
3. What are the endogenous influences and how do they impact the power relations?

1.4 Research Methodology

1.4.1 Research Strategy: Qualitative Research and Multiple Case Study Approach

This book adopts a qualitative approach because qualitative methods are particularly well-suited to identifying influences and power dynamics among affected actors in highly context-dependent settings (Creswell, 2013; Yin, 2013; Zhang et al., 2016). Building on this methodological orientation, the research is grounded in a social constructivist paradigm to examine how participatory policies shape local power relations in the urban redevelopment process, as shown in Table 1.2.

Table 1.2 Social constructivism research paradigm

Paradigm	Ontological Belief	Epistemological Belief	Axiological Belief	Methodological Belief
Social constructivism	Realities are constructed through experiences and interactions with others.	Reality is co-constructed between the researcher and the researched.	Individual values are honoured and negotiated among individuals.	An inductive process of research.

Source Adapted from Lincoln et al. (2011) and Creswell (2013)

From an ontological perspective, power relations are not merely imprinted on individuals but are constituted through mutual interactions among social actors (Creswell, 2013). These interactions construct the social realities that this research seeks to investigate. Epistemologically and axiologically, this book acknowledges the active role of the researcher in knowledge production and assumes that power relations can be examined through the analysis of negotiations among involved stakeholders in the decision-making process. Finally, based on the assumption that the influence of participatory redevelopment policies on power relations is manifested through interactions among stakeholders, an inductive research strategy is adopted to explore and explain these power relations in a theory-building manner.

This research is aiming to investigate the power relations among different stakeholders in participatory urban redevelopment, which is a highly complex, multifaceted and geographically diverse process. It requires a prolonged trace of empirical projects, rather than mere frequency or incidence. Given the contemporary and explorative nature of this research, the case study approach has been employed to address the research questions effectively.

Existing studies on power relations have predominantly relied on single-case designs, which tend to limit the ability to capture the systemic and differentiated patterns of power relations across diverse participatory redevelopment practices in China. To address this limitation, this research employs a multiple case study strategy to provide a more robust and comprehensive examination of power relations. Three cases are purposively selected to represent contrasting redevelopment modes—government-led, developer-led and owner-led projects. Case selection follows a replication logic to ensure comparability across cases. Through comparative analysis, this study seeks to identify not only shared patterns but also context-specific variations in the evolution of power relations in urban redevelopment.

1.4.2 Research Scope

This book adopts a purposive sampling strategy to select cases and interviewees in a strategic manner rather than on a random basis, in order to address the research questions. The purposive sampling process follows a sequential logic, beginning with the selection of the city and research scope and proceeding to case selection.

I. Case City: Why Shenzhen?

Shenzhen is selected as the study city for this research because it was the first city in China to initiate and implement participatory urban redevelopment policies (Hu & Li, 2025; Li & Liu, 2022; Li et al., 2019; 2021). Located in Guangdong Province and bordering Hong Kong Special Administrative Region, Shenzhen covers an area of approximately 1991 km^2. In the late 1970s, in response to the systemic inefficiencies and misallocation of resources under the centrally planned economy, the Chinese central state launched market-oriented reforms aimed at granting greater autonomy to local governments (Brandt & Rawski, 2008). Owing to its geographical proximity to Hong Kong, Shenzhen was designated as the first Special Economic Zone (SEZ) to pilot these reforms and soon emerged as a key testing ground for policy experimentation and economic transformation in China (Ng, 2003).

Since its establishment as an SEZ, Shenzhen has undergone a dramatic transformation from a small border town into a major metropolis over the past four decades, a process widely regarded as a "miracle" of urban development. During the 1980s, large numbers of labour-intensive industries relocated from Hong Kong to Shenzhen, fostering an export-oriented economy driven by market forces. Continuous capital accumulation under this development model enabled Shenzhen to attract advanced technological, managerial and financial resources from developed economies, facilitating industrial upgrading (Zhang, 2012). As a result, Shenzhen has gradually evolved into a national centre for high-tech industries, biotechnology and creative sectors (Chen et al., 2014). Globally recognized firms such as Huawei, Tencent, ZTE and DJI, together with more than three million registered businesses, generate substantial tax revenues for the municipal government (He, 2018). With strong fiscal capacity, the local government has invested heavily in infrastructure to cultivate a favourable business environment. The establishment of an international airport, three major ports and the Shenzhen Stock Exchange has further consolidated the city's role as a major logistics and financial hub in South China. Shenzhen's gross domestic product (GDP) increased from 270 million yuan in 1980 to 2.2 trillion yuan in 2018 (Wang, 2018), while its resident population grew from approximately 300,000 in 1979 to over 21 million in 2017 (Urban Planning China, 2018). These processes of marketization have positioned Shenzhen as one of China's first-tier cities and a leading forerunner of the country's capitalist transformation (He, 2018).

Rapid economic growth has been accompanied by extensive urban expansion. The area of construction land in Shenzhen increased from 2.81 km^2 in 1979 to 968 km^2 in 2014—an expansion of more than 300 times (Qian et al., 2015). With construction land occupying nearly half of the city's total area, environmental sustainability has been increasingly threatened (Chen et al., 2016). Rapid urbanization has also resulted in inefficient land use and substantial underutilized land within the built-up area (Shi & Fan, 2011), posing significant challenges to sustainable development. Notably, compared with other Chinese megacities, land conveyance fees account for a relatively small share of Shenzhen's fiscal revenue (Feng, 2019), indicating that the municipal government is less dependent on land auctions as a primary revenue source due to its robust market economy and business activity. These conditions jointly created strong incentives for a strategic shift towards urban redevelopment.

Confronted with acute land scarcity and declining land-use efficiency, Shenzhen was the first Chinese city to delineate an Urban Growth Boundary in 2005 (Qian et al., 2015). An ecological protection line was subsequently introduced, designating nearly half of the city's territory as protected areas where urban development is prohibited. With only around 3% of land remaining available for future development, the concept of *Cun Liang Gui Hua* was formally incorporated into the 2007 master plan (Qian et al., 2015). This planning approach emphasized urban redevelopment as the primary pathway for economic and social transformation, industrial restructuring and improved land-use efficiency (Liu, 2010). In 2009, Guangdong Province was selected as a pilot region for participatory urban redevelopment mechanisms. In alignment with provincial initiatives, Shenzhen subsequently formulated a set of participatory redevelopment policies to facilitate implementation. These policies significantly empowered existing land and property owners while constraining governmental authority, thereby fundamentally reshaping power structures in the urban redevelopment process.

Despite Shenzhen's prominent role as a national policy laboratory, existing empirical research on power relations in participatory redevelopment has largely focused on cities such as Guangzhou, Beijing and Xi'an. Although Shenzhen initiated participatory redevelopment reforms as early as 2004, systematic empirical analyses of institutional change and power relations in its redevelopment practices remain limited. This research therefore selects Shenzhen as the study city to address this empirical gap.

II. Case Scope: Why the Dilapidated Residential Area?

The demand for the dilapidated residential redevelopment is highly urgent in Shenzhen. Most of these dilapidated communities were built during the late 1980s and the early 1990s as welfare housing for migrant workers from all over the country (Ng, 2003). Following the housing reform of the 1990s, these housing units were privatized and sold by asset-holding work units, thereby becoming transferable on the housing market. Owing to outdated construction technologies and limited governmental financial resources at the time, many of these buildings were constructed with relatively poor quality. After completion, insufficient fiscal capacity further constrained long-term maintenance, resulting in deteriorating physical conditions, substandard living environments and growing dissatisfaction among residents.

Despite these deficiencies, most dilapidated residential areas are located within the inner ring of Shenzhen's built-up area, where land values are particularly high. The juxtaposition of poor building quality and low development density has thus led to significant underutilization of valuable urban land in the city centre.

This book focuses on dilapidated residential areas as its primary study scope for two reasons. First, the mechanisms governing residential redevelopment in Shenzhen have undergone profound transformation. The introduction of participatory redevelopment policies in 2009 stimulated a surge of redevelopment initiatives targeting dilapidated residential communities. Within a relatively short period, multiple redevelopment approaches—including government-led, developer-led and owner-led models—were experimented with by the local government in an effort to activate the premium value of centrally located land and to alleviate community grievances. The coexistence of these three redevelopment modes within dilapidated residential redevelopment provides an ideal empirical setting for a systematic analysis of power relations under participatory redevelopment.

Second, dilapidated residential areas—characterized by clearly defined property rights, relatively weak social associations and premium locations—present a distinctive and underexplored context for examining local-level power relations. Shenzhen's participatory redevelopment policy identifies four categories of underutilized land for redevelopment: urban villages, old industrial areas, old towns and dilapidated residential areas. Existing studies have largely concentrated on power relations in the redevelopment of old towns (e.g. Cheng, 2012; Yang & Chang, 2007) and urban villages (e.g. Chung, 2013; Li & Li, 2011), while empirical research on dilapidated residential redevelopment remains limited.

In dilapidated residential areas, property rights are legally defined yet fragmented among individual households. This clear and legitimate ownership structure substantially empowers property owners in negotiations with local governments and private developers, generating patterns of power deployment distinct from those observed in old towns and urban villages. Moreover, social relations within these commercial residential communities are generally weaker. Residents in old towns often share long-term co-residence and dense social networks, while urban villages are typically organized through village collectives that foster strong social cohesion. By contrast, property owners in dilapidated residential areas usually lack collective organizational structures, resulting in comparatively weak social ties. The combined differences in property regimes, social networks and physical conditions give rise to distinct stakeholder constellations and power configurations. Dilapidated residential areas therefore offer a particularly suitable context for examining how and why different interaction patterns among participants emerge and how these interactions ultimately shape the outcomes of participatory redevelopment processes.

III. Case Selection: why Ludancun, Mutoulong and Nanhuacun?

Three information-rich cases of dilapidated residential redevelopment are selected to investigate power relations in participatory processes: one government-led case (Ludancun), one developer-led case (Mutoulong) and one owner-led case (Nanhuacun). All three cases are located in the central area of Shenzhen (within the Guannei area, as shown in Fig. 1.2) and share highly similar development backgrounds and physical characteristics (see Table 1.3). They were constructed during a similar period by the municipal government as welfare housing for migrant workers. The communities exhibit comparable development densities and identical building typologies.

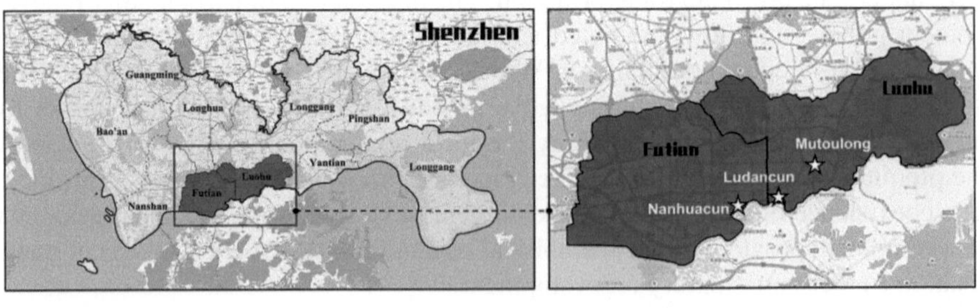

Fig. 1.2 Location of three cases

Table 1.3 Information of three dilapidated residential cases

	Case1	Case2	Case3
Name	Ludancun	Mutoulong	Nanhuacun
Area	72,450 m^2	83,234 m^2	161,900 m^2
Previous FAR	1.51	1.55	1.45
Building type	7-story building without elevator	7-story building without elevator	7-story building without elevator
Construction time	1988	1986	1986
Household numbers	1300	1341	2524
Residential area status	Originally constructed as welfare housing for employees of major work units, these buildings have since deteriorated and are now in a state of disrepair.		
Land use	Residential	Residential	Residential
Property ownership	Fragmentedly owned by individual households	Fragmentedly owned by individual households	Fragmentedly owned by individual households
Redevelopment status	Completion	Severe delay in property assembly	Without project approval
Redevelopment mode	Land preparedness (government-led)	Urban renewal (developer-led)	Urban renewal (owner-led)

Due to inferior construction materials and techniques, combined with decades of inadequate maintenance, all three communities have deteriorated into dilapidated housing areas characterized by poor living conditions and outdated infrastructure. Despite these similarities, the cases have followed three distinct redevelopment trajectories as a result of differences in social composition, degrees of physical deterioration and configurations of power relations, which in turn produced divergent redevelopment processes and outcomes.

Ludancun represents a government-led residential redevelopment case. The community experienced severe deterioration in living conditions, prompting repeated petitions from residents. In response, the local government officially initiated the redevelopment of Ludancun in 2003. However, in the absence of an appropriate policy framework to safeguard residents' interests, affected residents persistently resisted the government's initial proposals. Prolonged negotiations failed to produce consensus. Following the introduction of participatory urban redevelopment policies in 2009, the local government adopted a government-led redevelopment mechanism that incorporated greater interest-sharing arrangements with residents. This institutional shift incentivized the majority of residents to cooperate, thereby facilitating the successful completion of the Ludancun redevelopment project. Notably, Ludancun became the first successfully completed residential redevelopment project in Shenzhen and remains the only government-led residential redevelopment case in the city.

Mutoulong represents a developer-led residential redevelopment case initiated after the establishment of market-oriented redevelopment policies in 2009. Attracted by the community's premium location and large scale, a private developer entered the project and secured redevelopment support from a majority of affected residents. The developer subsequently applied for approval under the market-led redevelopment mechanism, and Mutoulong was designated as an experimental developer-led dilapidated residential redevelopment case. However, due to external policy changes and internal conflicts among key stakeholders—including the developer, cooperative residents and non-cooperative residents—negotiations have remained unresolved for more than nine years. As a result, Mutoulong, along with seven other experimental cases, has experienced prolonged stagnation. In response, the local government suspended the approval of market-led redevelopment for future residential redevelopment projects in 2012.

Nanhuacun represents an owner-led residential redevelopment case. Owing to its large scale and the relatively low level of initial consensus among affected residents, Nanhuacun was not included among the experimental cases for developer-led redevelopment. Dissatisfied with persistently poor living conditions, residents—together with a real estate agent—began to pursue an owner-led redevelopment agenda under the participatory redevelopment framework, drawing on the community's historical experience of successful self-redevelopment three decades earlier. However, in response to severe delays in several developer-led redevelopment projects, the local government suspended the approval of all dilapidated residential redevelopment proposals. As a result, despite more than seven years of sustained efforts, residents in Nanhuacun have been unable to secure governmental approval for redevelopment and continue to live in deteriorating housing conditions.

Collectively, the three cases represent the three principal redevelopment approaches—government-led, developer-led and owner-led—and illustrate three distinct patterns of power relations among local governments, developers and households. Juxtaposing these cases not only captures the coexistence of multiple redevelopment modes but also reveals the evolutionary transformation of residential redevelopment mechanisms, thereby providing a comprehensive picture of participatory governance in Shenzhen (Fig. 1.3).

In addition, accessibility to interviewees constitutes a key consideration in case selection. In Ludancun, many residents are to be relocated on-site following the completion of redevelopment, enabling access to a relatively large and representative sample of affected residents. In Mutoulong and Nanhuacun, redevelopment projects remain stalled, and a proportion of residents continue to reside on site, making them readily accessible for interviews. Furthermore, social networks among remaining residents facilitate efficient contact with relocated residents. Based on considerations of interview accessibility and case comparability, Ludancun, Mutoulong and Nanhuacun are selected as the three cases for examining power relations in participatory redevelopment in Shenzhen.

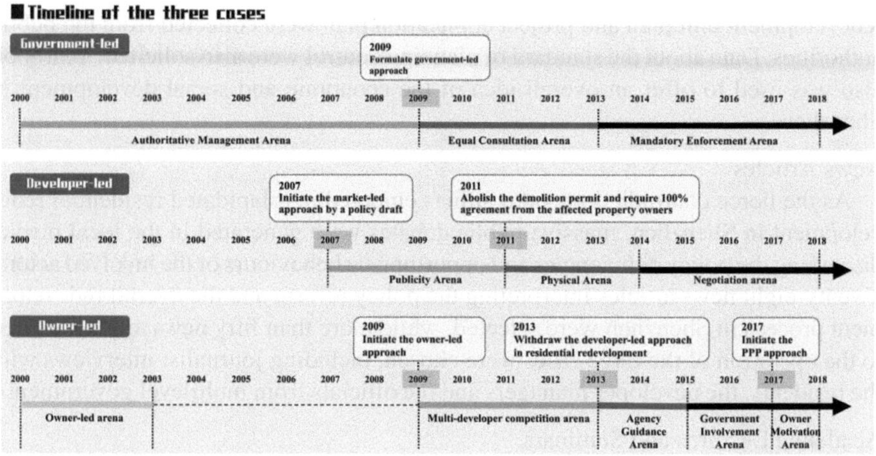

Fig. 1.3 Timeline of three cases

1.4.3 Data Collection

In order to fulfil a comprehensive understanding of the cases, this research uses different forms of qualitative data from six sources, including laws, policies and urban plans, news articles, social media such as online forums and blogs, in-depth interviews, field observation and academic lectures and seminars. By combining various empirical materials, researcher triangulates the data to increase the credibility and to overcome the weakness and problems which come from the single method.

I. Collection of Secondary Data

To achieve a preliminary understanding of the cases, this research uses different forms of secondary qualitative data from four sources, including governmental documents, news articles, academic lectures and seminars and social media such as online forums and blogs.

Governmental Documents

Since one of the main focuses of this research is to investigate the relationships between changing policies and the power relations in local redevelopment practice, official documents of different levels of hierarchy such as laws, policies, urban plans and yearbooks are considered as the primary sources of second-hand data. They set a framework for understanding the institutional background of urban development and redevelopment in Shenzhen, China.

More than 60 governmental documents are collected from the central government, the Shenzhen municipal government, land and planning bureau, urban redevelopment bureau and other urban development institutes, including property right law, land management policies, urban planning policies and participatory redevelopment policies from national, provincial to local level. Urban redevelopment schemes, urban redevelopment unit plan and project application plan were collected from the public authorities. Data about the standard of planning control were also solicited. Yearbook also was used to offer an overall idea of the economic and social development in Shenzhen.

News Articles

As the fierce contradiction and conflicts emerged in dilapidated residential redevelopment in Shenzhen, massive public debates were generated in the local media, discussing the policy deficiencies and opportunistic behaviours of the involved actors. Twenty-eight news articles interpreting the policy reform and the overall redevelopment process in Shenzhen were selected, while more than fifty news articles related to the operation of the three cases were chosen, including journalist interviews with the residents, the developer managers and the officials from multilevel governments.

Academic Lectures and Seminars

The researcher also attended two lectures and one seminar related to the participatory redevelopment policy reform in Shenzhen. The two lectures were held by the Shenzhen government to promulgate its newest policies to urban planners and developers. The seminar was set up by the Shenzhen real estate institute to introduce the new urban redevelopment mechanism to scholars from Chongqing University. These academic lectures and seminars improve the researcher's understanding of Shenzhen's participatory redevelopment policies.

Social Media

New forms of data have collected from the online social media, such as Sina blogs, Shenzhen BBC online forum and QQ groups. For the case of Ludancun, the residential committee used an online blog to publicize information and seek for property owners' opinions since December 2009. Till 12 July 2017, there are 118 blogs in total, which can be seen as partly detailed records of the participatory redevelopment process for Ludancun. Some leading uncooperative property owners in Mutoulong posted their experiences on the Shenzhen BBC online forum. These posts provided detailed records about the progress of Mutoulong redevelopment, including the requirement of the uncooperative residents, the opportunistic behaviours of the developer, the conflicts between cooperative and uncooperative residents and so forth. In the case of Nanhuacun, property owners posted public letters about their poor living conditions and petitions for redevelopment through the online forum, which provided evidence about residents' requirement. The real estate agent cooperated with some active property owners to organize a QQ group, which worked as a platform to share information and form solidarity among the fragmented property owners. Documents were accessible about redevelopment supports, compensation contracts, meeting records and so forth. Data from these social media offered specific evidence of the strategies and mobilized actions of the involved stakeholders in the participatory redevelopment process.

The collection of secondary data was considered as preparation work for fieldwork. It began when the research questions were decided. The purposes of this work were to identify the participatory redevelopment policies, discover the key stakeholders and assess the main redevelopment process and prepare for interviews and field reconnaissance. The first step was to get familiar with the general redevelopment procedure and governing process in Shenzhen. I made full use of the internet resources—keywords were used to search all the relevant policies and regulations at different levels, such as "*Cun Liang Gui Hua*", "redevelopment", "participatory redevelopment". The governmental documents about policies, regulations and redevelopment unit plans were easily obtained. The researcher spent one month on these data and had acquittance of a full understanding of the participatory redevelopment procedure and the roles of the main stakeholders.

Secondly, keywords such as "Ludancun", "Mutoulong", "Nanhuacun" and "redevelopment" were utilized for searching relevant news, posts and blogs. These internet resources helped the researcher get sufficient context about the three cases and residents' complaints about the redevelopment projects. Based on a more in-depth understanding of the three cases, the researchers organized the interview questions in a more detailed and efficient way.

The secondary data found at this stage was very useful but had some limitations, especially in terms of the material found on the news and social media. Firstly, some key data identified as required was not available either online or offline. For example, some actors were mentioned to take some mobilized actions in the redevelopment process. The motivations and the objectives behind the actions were inaccessible on the internet. Secondly, the data found were often biassed. Most news articles were very positive about the redevelopment programs, probably due to the censorship regulations. Few complaints, arguments and conflicts, therefore, had been reported in public media. Publicity materials from the developer headquarters had the same problem. Other information sources found online, including articles and posts from social network sites, appeared to be written by residents. The information mostly concerned the adverse effects of redevelopment programs, with residents making complaints, doubting the project and reporting issues they had witnessed and suffered in their communities. It would have been helpful if the researcher could have interviewed the authors of these pieces, but unfortunately, it was challenging to contact them. Thirdly, the volume of data sources was limited, with online data, especially news items, comprising a large proportion of all the secondary data. Although data collected at the preparation stage had these limitations, it was useful and sufficient to understand the case study contextual features.

II. Field Observation

Data collected from field observation are composed of two compositions: notes and photos. Notes are mainly relevant with descriptions about existing residents living style, living environment, residents' intra-relations and mobilized actions on site. Photos were taken to document residents a physical living environment of the dilapidated residential areas, site conditions and also record some posters and noticeboards. These data are used to corroborate evidence with interview data to depict a comprehensive picture.

Eleven field observations were conducted during fieldwork to collect information regarding urban built environment changes and residents' living environment. Before the initiation of the official interviews, field observation had been conducted for each case. Photos about the three cases were taken to record their physical attributes and several casual conversations with residents were carried out. Notes were also taken about existing residents' living style and their living environment. After on-site interviews, a preliminary understanding of the relationship among the main stakeholders has been developed. Several field trips were undertaken to observe the interrelationship between the existing residents and the way they communicated with each other.

III. Semi-Structured Interviews

There are 34 interviews, covering key informants including officials from (1) municipal government, (2) district government, (3) the land and planning bureau, and (4) the planning and design institute, and representatives of (5) the management team of real estate company, (6) lawyers, (7) the street office (Jiedaoban), (8) the work station (Gongzuozhan) and (9) property owners. The interviews varied, in terms of the length, being audio recorded or not, in a group or not. All formal interviews were summarized in the appendix.

Since topics such as "power", "the role of the government" and "mobilized actions of residents" are highly sensitive when doing fieldwork in China, semi-structured interviews are proposed as the main method for this study. This is because semi-structured interviews can collect data in a more in-depth understanding of the research topic and provide more appropriate freedom for respondents to discuss sensitive topics (Fylan, 2005; Louise Barriball & While, 1994).

Based on the secondary data and casual conversations with residents, relevant stakeholders and their administrative relationships were identified. Additionally, a list of critical questions was developed, in accordance with research questions and second-hand data, to refer to during interviews. The guide was used as a checklist to direct the development of open-ended questions and to make sure that all the relevant points were covered in the interviews. On the basis of each point, many questions were asked accordingly. For example, reference "power relations in participatory redevelopment" led to questions such as "how did the participatory redevelopment policy influence your decisions and actions", "what role do you play in the participatory redevelopment process", "how other actors react to your strategies and actions" etc. The interview question list was provisional but not fixed. Once new key points were identified from interviews, the list was expanded; or once things turned out to be different from what had been expected from the interview guide, it was adapted to adjust to the real situation.

Two strategies were adopted in seeking of interviewees who are related to the research topics, including on site recruiting and snowballing. On site recruiting strategy was adopted initially to commence the interview. A small group of interviewees including cooperative and uncooperative residents, government officials and urban planners were accessible through site visiting in the three cases, relevant government departments and planning and design institutes. During the interview, these sampled participants proposed other participants who have had the experience or characteristics pertinent to the research, based on their networks. For example, one government cadre in the urban renewal office in Luohu District regarded this research was of high relevance and significance for the dilapidated residential redevelopment in Shenzhen. He contacted the project manager of Mutoulong redevelopment and arranged an interview between the researcher and the manager. Most interviews were recorded with the participant's permission.

Moreover, because this research is highly related to the topics which were considered as sensitive in China, such as power, interest, the role of the government, community participation and so on, some interviewees might fail to frankly express their feelings out of the concern about social security and privacy. To address this challenge, the researchers conducted a pilot study in the Ludancun, where informants held a more open and positive attitude because of the success of the project. Through the pilot study, both the interview questions and communicative skills were improved, which were able to benefit the following fieldwork (Suoninen & Jokinen, 2005). During the investigation in this research, the researcher avoided directly asking questions about power and interest distribution but to inquire about the role of each participant, the process of the redevelopment and the interactions between stakeholders in a neutral position. To protect the anonymity of participants, all interviewees are referred as pseudonym in this thesis. All materials such as audio recordings, interview notes, transcriptions and signed consent forms are stored in a locked cabinet and a password protected PC in the university office.

1.4.4 Data Analysis

This research adopts Creswell (2013)'s qualitative data analysis approach including the following steps: organizing the data, conducting a preliminary read-through of the database, coding and organizing themes, representing the data and forming an interpretation (Fig. 1.4).

Prior to the fieldwork, a substantial amount of secondary data—including government documents, news articles and social media content—was collected, organized into separate files and categorized. These materials provided a preliminary understanding of the participatory policy context and urban redevelopment practices in the three cases, serving as preparation for the fieldwork.

During the fieldwork, primary data were primarily gathered from informants through interviews, with audio recordings and field notes as the main sources. Audio recordings were transcribed immediately after each interview to minimize potential issues such as poor recording quality, background noise or memory lapses over time. Any biases or gaps in information were addressed during the transcription process. The transcribed texts, along with documents, observation notes and visual data (images) derived from audiovisual materials and field observations, were systematically organized into computer files. By the conclusion of the field research, transcription was nearly complete.

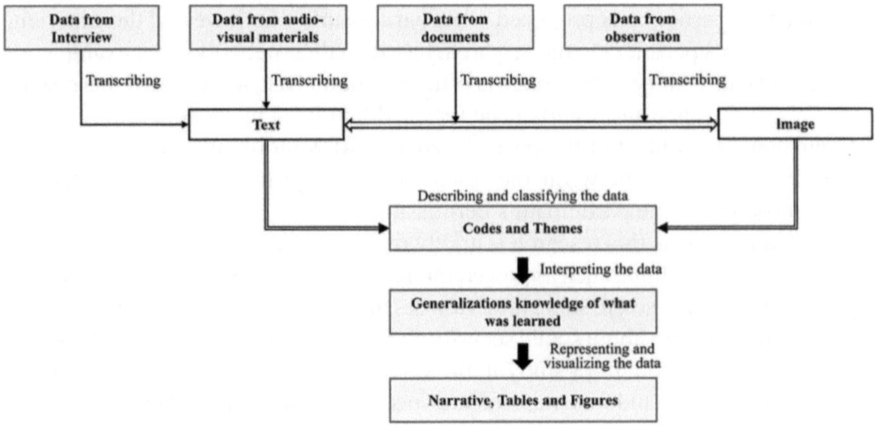

Fig. 1.4 Flow of Qualitative Data Analysis

The final combined dataset—including interview transcripts, news reports, social media content and observation notes—comprised approximately 80,000 words in Chinese. These transcripts were carefully reviewed multiple times to ensure the researcher's familiarity with the content prior to formal analysis.

Following transcription and organization, the raw data were coded and sorted into various thematic categories. Guided by the analytical framework, the researcher developed a reference table (see Table 1.4) to facilitate coding. This table outlined initial themes for data classification but was revised and refined flexibly throughout the analytical process. Categories and subcategories were conceptualized at the outset. The transcripts were then read and coded selectively, either line-by-line or by thematic units. It was not necessary to code every word, sentence or paragraph; for instance, text irrelevant to the research questions or repeated and redundant content was excluded from coding. After completing the initial coding, all codes were reviewed for refinement, including further conceptualization and theorization of code names, merging of similar codes and deletion of unrelated or insignificant codes.

Table 1.4 Initial framework for data analysis

Category	Sub-category
Exogenous influences	Policy reform Administrative intervention
Power relations	Participants Aims and objectives Power and its resources Strategies and actions Impacts of the power deployment
Outcome	Spatial outcomes Benefit and cost Social and environmental impacts

Once the coding process was completed, the analysis proceeded to the sorting stage. Excerpts assigned to the same code were extracted and organized into corresponding folders labelled with the code titles. This operation was facilitated using computer-based word processing software. The result of the sorting process was a set of folders, each containing excerpts corresponding to a particular code. Through coding and sorting, the textual and visual data were systematically divided into focused thematic units, while irrelevant or non-informative data were identified and discarded (Wolcott, 1994).

According to Lincoln and Guba (1985), data analysis should extend beyond the identification of themes and codes, situating tentative explanations within the broader dataset and linking them to existing research literature. To achieve this, the present study employed a process of local integration, interpreting the data to generate generalized knowledge as defined in the power framework. For example, from an interview, the researcher observed that developers possessed financial and technological resources to carry out participatory processes, reflecting power derived from allocative resources.

The final step involved inclusive integration, in which the analytical framework was employed to synthesize the collection of coded excerpts into a coherent narrative, progressing logically from one thematic area to the next and ultimately leading to general conclusions. The researcher examined the analysis from the perspective of local integration to identify interrelationships among themes. For example, upon completing the power relations analysis across the three cases, the findings were related to existing literature to generate insights into the evolution of governance systems in Chinese urban redevelopment.

1.5 Structure of the Book

The book is organized into eight chapters. Chapter 1 outlines the research problem and gaps, research objectives and questions, the empirical literature review, the research methodology and the overall structure of the book. The empirical review is organized around three stages of urban redevelopment: state-dominated redevelopment in the socialist period, growth-coalition–dominated redevelopment during the market-oriented reform period and participatory redevelopment in the participatory period. The chapter then explains the rationale for adopting a multiple case study approach and a purposive sampling strategy for recruiting informants for semi-structured interviews and site observations. The selection of Shenzhen and its dilapidated residential areas is further justified. Finally, the chapter details the data collection process, including secondary data analysis, 11 site observations and 34 semi-structured interviews.

Chapter 2 develops the analytical framework through a systematic review of the literature on power, power relations and institutional analysis. The review is structured around three themes: (1) debates on power relations in urban development; (2) the conceptualization of power, power resources and power relations and (3) institutions, rules and institutional analysis, through which the theoretical gaps in existing discussions of power relations are identified. Building on this review, the chapter clarifies the key concepts of power, power resources and power relations and establishes the power arena framework.

Chapter 3 examines the institutional reforms associated with participatory redevelopment policies, thereby providing the institutional context for the thesis. It begins by outlining the setting, processes and consequences of rapid urban development over the past three decades. The chapter then elaborates on the institutional arrangements established by central and provincial governments to implement participatory redevelopment policies. This is followed by a review and analysis of local measures adopted by the Shenzhen municipal government to facilitate participatory redevelopment. Finally, the chapter illustrates the diverse redevelopment approaches for dilapidated residential areas in Shenzhen, including government-led, developer-led and owner-led.

Chapter 4 investigates the power relations among the main stakeholders—the local government, developers and affected residents—in a government-led participatory redevelopment project. The chapter begins with a profile of the case study, followed by an examination of social practices and power relations among the stakeholders. The impacts of both exogenous and endogenous influences are then analysed. The chapter further presents and evaluates the spatial and social outcomes of the project, with particular attention to issues of equity, adaptability and transaction costs.

Chapters 5 and 6 apply the same analytical structure as Chap. 4 but focus on projects adopting different redevelopment approaches. Chapter 5 examines a developer-led participatory redevelopment project, while Chap. 6 analyses an owner-led participatory redevelopment project.

Chapter 7 provides an in-depth discussion of power relations in the participatory redevelopment process. It begins by examining changes in power relations in urban redevelopment in China. The chapter then analyses shifts in exogenous and endogenous influences and their impacts on power relations. Building on this analysis, the chapter connects discussions of power, power relations and institutions to elucidate how the reciprocal interplay between structure and agency shapes power relations in participatory redevelopment.

Chapter 8 concludes the book by summarizing the main findings and presenting the overall conclusions of the study. It also discusses research limitations and offers directions for future research as well as suggestions for policy innovation.

References

Balding, C. (2019). What's causing China's economic slowdown. *Foreign Affairs*. https://www.foreignaffairs.com/articles/china/2019-03-11/whats-causing-chinas-economic-slowdown

Bevir, M. (1999). Foucault, power, and institutions. *Political Studies, 47*(2), 345–359.

Brandt, L., & Rawski, T. G. (2008). *China's great economic transformation.* Cambridge University Press.

Bunker, R., Gleeson, B., Holloway, D., & Randolph, B. (2002). The local impacts of urban consolidation in Sydney. *Urban Policy and Research, 20*(2), 143–167. https://doi.org/10.1080/08111140220144461

Chen, H., Wang, X., & Guo, Z. (2015). The transition of China's urban planning: Thoughts on increment planning, inventory planning and reduction planning. Modern Urban *Research*(09), 44–48.

Chen, J., Chang, K.-t., Karacsonyi, D., & Zhang, X. (2014). Comparing urban land expansion and its driving factors in Shenzhen and Dongguan, China. *Habitat International, 43*, 61–71.

Chen, Y., Chen, Z., Xu, G., & Tian, Z. (2016). Built-up land efficiency in urban China: Insights from the general land use plan (2006–2020). *Habitat International, 51*, 31–38. https://doi.org/10.1016/j.habitatint.2015.10.014

Cheng, Z. (2012). The changing and different patterns of urban redevelopment in China: A study of three inner-city neighborhoods. *Community Development, 43*(4), 430–450. https://doi.org/10.1080/15575330.2012.711763

Chung, H. (2013). The spatial dimension of negotiated power relations and social justice in the redevelopment of villages-in-the-city in China. *Environment and Planning A, 45*(10), 2459–2476. https://doi.org/10.1068/a45416

Cohen, I. J. (1989). *Structuration theory: Anthony Giddens and the constitution of social life*. Macmillan.

Creswell, J. W. (2013). *Qualitative inquiry and research design: Choosing among five approaches*. Sage.

Dahl, R. A. (1961). *Who governs?: Democracy and power in an American city*. Yale University Press.

Dowall, D. E. (1994). Urban residential redevelopment in the People's Republic of China. *Urban Studies, 31*(9), 1497–1516.

Fang, K., & Zhang, Y. (2003). Plan and market mismatch: Urban redevelopment in Beijing during a period of transition. *Asia Pacific Viewpoint, 44*(2), 149–162.

Feng, K. (2019). The land sales ranking in China: Hangzhou top and Shenzhen bottom. *Guomin Jinglue.*. https://cj.sina.com.cn/articles/view/1632067682/6147606201900fbcb?from=finance

Foucault, M. (1982). *The subject and power. Critical inquiry, 8*(4), 777–795.

Foucault, M. (1998). *The will to knowledge: The history of sexuality, vol. 1* (Trans. R. Hurley). Penguin.

Fylan, F. (2005). Semi-structured interviewing. In J. Miles & P. Gilbert (Eds.), *A handbook of research methods for clinical and health psychology* (pp. 65–78). Oxford University Press.

Geoffrey, M. H., & Hodgson, R. (2006). What are institutions? *Journal of Economic Issues, 40*(1), 1–25.

Giddens, A. (1979a). *Central problems in social theory: Action, structure, and contradiction in social analysis* (Vol. 241). University of California Press.

Giddens, A. (1984). *The constitution of society: Outline of the theory of structuration*. University of California Press.

Gregson, N. (1989). On the (ir) relevance of structuration theory to empirical research. In *Social theory of modern societies: Anthony Giddens and his critics* (pp. 235–248). Cambridge University Press.

Han, S. S. (1994). *Controlled urbanization in China, 1949–1989* (Theses). Department of Geography)/Simon Fraser University.

Han, S. S. (2000). Shanghai between state and market in urban transformation. *Urban Studies, 37*(11), 2091–2112. https://doi.org/10.1080/713707226

Han, S. S., & O'Connor, K. (2008). Urban consolidation and house prices: A case study of Melbourne 1990–2004. *GeoJournal, 73*(4), 285–295.

Han, S. S., & Wang, Y. (2003). The institutional structure of a property market in inland China: Chongqing. *Urban Studies, 40*(1), 91–112.

Hayter, R., & Han, S. S. (1998). Reflections on China's open policy towards foreign direct investment. *Regional Studies, 32*(1), 1–16.

He, H. (2018). Shenzhen surpasses US$338 billion GDP mark in 2017, beats Hong Kong and Singapore's growth. *South China Morning Post.*. https://www.scmp.com/news/china/economy/article/2128310/shenzhen-88-cent-hi-tech-growth-roll-hit-y2tr-2017

He, S., & Wu, F. (2005). Property-led redevelopment in post-reform China: A case study of Xintiandi redevelopment project in Shanghai. *Journal of Urban Affairs, 27*(1), 1–23. https://doi.org/10.1111/j.0735-2166.2005.00222.x

He, S., & Wu, F. (2009). China's emerging neoliberal urbanism: Perspectives from urban redevelopment. *Antipode, 41*(2), 282–304. https://doi.org/10.1111/j.1467-8330.2009.00673.x

Healey, P., Cars, G., Madanipour, A., & De Magalhaes, C. (2002). Transforming governance, institutionalist analysis and institutional capacity. In *Urban governance, institutional capacity and social Milieux* (pp. 20–42). Routledge.

Hsing, Y. (2010). *The great urban transformation: Politics of land and property in China.* Oxford University Press. doi:https://doi.org/10.1093/acprof:oso/9780199568048.001.0001

Hu, W., & Li, X. (2025). Variegated state entrepreneurialism in China: Land commodification and housing provision in Shenzhen and Chongqing. *Urban Policy and Research*, 1–15. https://doi.org/10.1080/08111146.2025.2472057.

Hunter, F. (1953). *Community power structure a study of decision makers.* The University of North Carolina Press.

Jiang, X. (2019).*Economic structure is optimizing.* ChinaDaily.com.cn. http://www.chinadaily.com.cn/a/201901/22/WS5c4652bda3106c65c34e5b31.html

Li, B., & Liu, C. (2018). *Emerging selective regimes in a fragmented authoritarian environment: The 'three old redevelopment' policy in Guangzhou, China from 2009 to 2014. Urban Studies, 55*(7), 1400–1419. https://www.jstor.org/stable/26510376

Li, B., Tong, D., Wu, Y., & Li, G. (2019). Government-backed 'laundering of the grey'in upgrading urban village properties: Ningmeng Apartment Project in Shuiwei Village, Shenzhen, China. *Progress in Planning, 146*, 100436.

Li, L., & Li, X. (2011). Redevelopment of urban villages in Shenzhen, China—An analysis of power relations and urban coalitions. *Habitat International, 35*(3), 426–434. https://doi.org/10.1016/j.habitatint.2010.12.001

Li, X., Li, B., & Jiang, W. (2024). State-led versus market-led: How institutional arrangements impact collaborative governance in participatory urban regeneration in China. *Habitat International, 150*, 103134. https://doi.org/10.1016/j.habitatint.2024.103134

Li, X., & Liu, T. (2022). Nomocracy versus teleocracy: Comparing participatory urban redevelopment in Shenzhen and Kashgar. *Journal of Urban Affairs*, 1–21. https://doi.org/10.1080/07352166.2022.2053779

Li, X., Wu, H., & Han, S. S. (2021). Institutional Deficiencies in Market-Led Residential Redevelopment in Shenzhen, China. *Journal of Planning Education and Research*, 0739456X211051779. https://doi.org/10.1177/0739456X211051779.

Lin, G. C. S. (2015). The redevelopment of China's construction land: Practising land property rights in cities through renewals. *The China Quarterly, 224*, 865–887.

Lin, J., & Wu, J. (2015). Methods and strategies on storage developable land planning: The planning practice of Guanggang new town. *Huazhong Architecture, 02*, 96–102.

Lincoln, Y. S., & Guba, E. G. (1985). *Naturalistic inquiry* (Vol. 75). Sage.

Lincoln, Y. S., Lynham, S. A., & Guba, E. G. (2011). *Paradigmatic controversies, contradictions, and emerging confluences, revisited* (Vol. 4). Sage.

Liu, X. (2010). Urban renewal unit exploration and practice: Using Shenzhen as an example. *Planners, 26*(11), 66–69 [in Chinese].

Logan, J. R., & Molotch, H. (1987). *Urban fortunes: The political economy of place.* University of California Press.

Louise Barriball, K., & While, A. (1994). Collecting data using a semi-structured interview: A discussion paper. *Journal of Advanced Nursing, 19*(2), 328–335.

McCrea, R., & Walters, P. (2012). Impacts of urban consolidation on urban liveability: Comparing an inner and outer suburb in Brisbane, Australia. *Housing, Theory and Society, 29*(2), 190–206. https://doi.org/10.1080/14036096.2011.641261

Molotch, H. (1976). The city as a growth machine: Toward a political economy of place. *American Journal of Sociology, 82*, 309–332.

Moos, A., & Dear, M. J. (1986). Structuration theory in urban analysis: 1. Theoretical exegesis. *Environment and Planning A, 18*(2), 231–252.

Xinhua News. (2017). China's economy steadies on structural upgrade. *Chinadaily.com.cn*. http://www.chinadaily.com.cn/business/2017-06/14/content_29744422.htm

Ng, M. K. (2003). Shenzhen. *Cities, 20*(6), 429–441.

Ostrom, E. (1986). An agenda for the study of institutions. *Public Choice, 48*(1), 3–25.

Ostrom, E. (2005). *Understanding institutional diversity*. Princeton University Press.

Phipps, A. G. (2000). A structurationist interpretation of community activism during school closures. *Environment and Planning A, 32*(10), 1807–1823.

Qian, J., Peng, Y., Luo, C., Wu, C., & Du, Q. (2015). Urban land expansion and sustainable land use policy in Shenzhen: A case study of China's rapid urbanization. *Sustainability, 8*(1), 16. https://doi.org/10.3390/su8010016

Schoon, S. (2014). Chinese strategies of experimental governance. The underlying forces influencing urban restructuring in the Pearl River Delta. *Cities, 41*, 194–199. https://doi.org/10.1016/j.cities.2014.01.008

Sewell Jr., W. H. (1992). A theory of structure: Duality, agency, and transformation. *American Journal of Sociology, 98*(1), 1–29.

Shi, A., & Fan, Z. (2011). Regulation on construction land scale from increment-led to revitalization-oriented. *City Planning Review, 35*(8), 88–96 [in Chinese].

Shin, H. B. (2009). Residential redevelopment and the entrepreneurial local state: The implications of Beijing's shifting emphasis on urban redevelopment policies. *Urban Studies, 46*(13), 2815–2839.

Stone, C. N. (1989). *Regime politics: Governing Atlanta, 1946–1988*. University Press of Kansas.

Stone, C. N. (1998). Regime analysis and the study of urban politics, a rejoinder. *Journal of Urban Affairs, 20*(3), 249–260.

Suoninen, E., & Jokinen, A. (2005). Persuasion in social work interviewing. *Qualitative Social Work, 4*(4), 469–487.

Tang, W. S. (1994). Urban land development under socialism: China between 1949 and 1977. *International Journal of Urban and Regional Research, 18*(3), 392–415.

Tian, L., & Yao, Z. (2018). From state-dominant to bottom-up redevelopment: Can institutional change facilitate urban and rural redevelopment in China. *Cities, 76*, 72–83.

Urban Planning China. (2018). How many resident population in Shenzhen? We use big data to answer! *Sohu News* [in Chinese].

Wang, F. (2018). *The total gross domestic product has been more than 2.2 trillion yuan in Shenzhen* (Economic report in 21st century).

Wolcott, H. F. (1994). *Transforming qualitative data: Description, analysis, and interpretation*. Sage.

Wu, F. (1996). Changes in the structure of public housing provision in urban China. *Urban Studies, 33*(9), 1601–1627.

Wu, F. (2004). Residential relocation under market-oriented redevelopment: The process and outcomes in urban China. *Geoforum, 35*(4), 453–470.

Wu, F., Xu, J., & Yeh, A. G.-O. (2006). *Urban development in post-reform China: State, market, and space*. Routledge.

Yang, Y.-R., & Chang, C.-H. (2007). An urban regeneration regime in China: A case study of urban redevelopment in Shanghai's Taipingqiao area. *Urban Studies, 44*(9), 1809–1826. https://doi.org/10.1080/00420980701507787

Ye, L. (2011). Urban regeneration in China: Policy, development, and issues. *Local Economy, 26*(5), 337–347. https://doi.org/10.1177/0269094211409117

Yeh, A. G., & Wu, F. L. (1996). The new land development process and urban development in Chinese cities. *International Journal of Urban and Regional Research, 20*(2), 330.
Yin, R. K. (2013). *Case study research: Design and methods.* Sage.
Zhai, B., & Ng, M. K. (2013). Urban regeneration and social capital in China: A case study of the drum tower Muslim District in Xi'an. *Cities, 35*, 14–25.
Zhang, J. (2012). From Hong Kong's capitalist fundamentals to Singapore's authoritarian governance: The policy mobility of neo-liberalising Shenzhen. *China. Urban Studies, 49*(13), 2853–2871. https://doi.org/10.1177/0042098012452455
Zhang, L., Chen, J., & Tochen, R. M. (2016). Shifts in governance modes in urban redevelopment: A case study of Beijing's Jiuxianqiao area. *Cities, 53*, 61–69. https://doi.org/10.1016/j.cities.2016.01.001
Zhang, T. (2002). Urban development and a socialist pro-growth coalition in Shanghai. *Urban Affairs Review, 37*(4), 475–499. https://doi.org/10.1177/10780870222185432
Zhang, X., Hu, J., Skitmore, M., & Leung, B. Y. (2013). Inner-city urban redevelopment in China metropolises and the emergence of gentrification: Case of Yuexiu, Guangzhou. *Journal of Urban Planning and Development, 140*(4), 05014004.
Zhao, P. (2011). Managing urban growth in a transforming China: Evidence from Beijing. *Land Use Policy, 28*(1), 96–109. https://doi.org/10.1016/j.landusepol.2010.05.004
Zhao, Y. (2014). Stock planning: Theory and practice. *Beijing City Planning and Construction Review, 4*, 153–156 [in Chinese].
Zhou, C., He, Y., Guo, Q., & Chen, W. (2014, May 29). *The historical records of Ludancun.* http://epaper.oeeee.com/epaper/H/html/2014-05/29/content_3248972.htm?div=-1.
Zhou, Z. (2014). Towards collaborative approach? Investigating the regeneration of urban village in Guangzhou, China. *Habitat International, 44*, 297–305.
Zhu, J. (1999). Local growth coalition: The context and implications of China's gradualist urban land reforms. *International Journal of Urban and Regional Research, 23*(3), 534–548.
Zhu, J. (2002). Urban development under ambiguous property rights: A case of China's transition economy. *International Journal of Urban and Regional Research, 26*(1), 41–57.
Zou, B. (2015). The transformation from greenfield-based planning to redevelopment planning: Theoretical analysis and practical strategies. *Urban Planning Forum, 225*(5), 12–19 [in Chinese].
Zou, B. (2017). Practices, effects and challenges of the inventory development pattern: The assessment and extended thoughts of urban renewal implementation in Shenzhen. *City Planning Review, 41*(1), 89–94 [in Chinese].

Chapter 2
Power Arena Framework: Power, Power Relations and Institutions

2.1 Introduction

This chapter presents the theoretical framework by organizing a theoretical review of the literature on power relations and institutions. Section 2.2 focuses on the current theories of power relation debates in urban development. Section 2.3 assesses the conceptualization of power, power resources and power relations based on the structuration theory and Foucault's power approach. Section 2.4 examines the utility of the institutionalist approach in understanding the institutional setting and outcomes. Section 2.5, based on the above theoretical discussions, establishes the theoretical framework of this book—power arena framework.

2.2 The Debates of Power Relations in Urban Development

Three main theses about power relations in urban development are critically reviewed in this section. They are (1) the community power debate, (2) the growth machine theory and (3) the regime theory. The community power debate recognized the influence of individual agents, some of whom had a more significant impact than others on the decision-making process. The growth machine theory emphasized the importance of individuals and interest groups in a broad political-economy context of urban development. The regime theory conceptualized the coalition of agents as a "regime", the formation of which was based on strict criteria. It also provided a new perspective that power can stem from social production. However, these theories were intrinsically positioned in the capitalist economy where liberal democracy was deeply rooted. Their suitability to the Chinese context is always in questions.

X. Li, *Participatory Redevelopment as Contested Arenas in China*,
https://doi.org/10.1007/978-981-95-9394-1_2

2.2.1 The Community Power Debate—Elite Theory and Pluralism

The community power debate began from Hunter's study of the governing structure in Atlanta (Hunter, 1953). In that study, Hunter applied reputational analysis by measuring the power of individuals according to their reputation for having that power. Hunter indicated that power was considered as the property of individuals and only a small set of policymakers were able to hold "power" and make all significant decisions in the city. Though a seminal study, Hunter ignored "positional feature" that different social roles and positions might endow diversified power to people (Harding, 1995). It was criticized that the dependence on reputations could not possibly prove individual's deployment of power in particular situation (Judge et al., 1995).

Hunter's study triggered the "community power debate" between elitist and pluralist theorists. Dahl (1961), holding a pluralist stand (Judge et al., 1995), adopted "decisional analysis" to examine decision-making process in several policy areas in New Haven, such as urban redevelopment, public education and political nominations. In his analysis, Dahl argued that power was never controlled by a single, coherent elite though it was impossible that power would equally distribute across all social groups (Dahl, 1961). Building upon Weber's definition of power[1], Dahl explained that A had power over B to the extent to which A could get B to do something which B would not otherwise do and the mobilization of resources was the utility of power in decision-making. In the analysis of three key areas in New Haven, Dahl (1961) found only a small number of the people had a direct influence on the decision-making while most citizens possessed a moderate degree of indirect influence. Dahl (1961) called the phenomenon the "dispersed inequalities" and explained that the unequal distribution of resources was the reason. Pluralist Polsby (1980) further argued that what mattered was not mere possession of resources but the processes of bargaining, negotiation, salesmanship, brokerage and leadership in mobilizing those resources.

The decisional analysis was criticized, in turn, by neo-elite theorists. Bachrach and Baratz (1962, 1970) argued that decisional analysis was not able to indicate the way policy agendas were set and failed to take into consideration of institutional bias. They concluded that the power used to make decisions depicted only one face of power (Bachrach & Baratz, 1962). In the second face, power also exercised by preventing issues from reaching the decision-making process, thus a decision not to make a decision led to "non-decision-making". However, pluralists challenged the "second face" of power on the grounds that invisible issues could not be studied and analysed as visible decisions (e.g. Waste, 1987). As a countermeasure, Polsby (1980) argued that non-decision was also a type of decision.

Based on the conceptualization of two-dimensional power, Steven (1974) added in the third dimension of power which arose out of socially structured and culturally patterned behaviour of groups, and practices of institutions. He redefined power

[1] Weber (1947) defines that power refers to the possibility that one actor who is of a certain position within a social relationship to carry out his own intention in spite of resistance.

in terms of interests—individual A exercised power over B when A affected B in a manner contrary to B's interests. The critical contribution of his work was the perception that power relations were shaped by structurally constituted social relations. However, his third-dimensional power was considered difficult to distinguish from the former two-dimensional powers (Haugaard, 2002). Another weakness was that it overlooked the relationship among structure, social knowledge and power.

The community power debate conceptualized and emphasized power in the decision-making process. By underlining the importance of human agency, the debate moved away from the idea that patterns of human activity were somehow determined by impersonal forces alone or were simply natural and explained by their functions. Scholars in this debate raised important issues about power relations from the perspective of "who benefited" from urban decision-making; they demonstrated that urban theory could be critical and normative as well as analytical and descriptive.

However, both elite theory and pluralist theory treated the cities they studied as if they existed in an institutional vacuum thus failing to consider the external influences and power configuration in a more generalized situation (Blokland, 2003; Wellman & Leighton, 1979). Further, it was argued that patterns of human action were not exclusively determined by impersonal (e.g., natural and functional) forces as decision-making processes were shaped by individual behaviours (Judge et al., 1995). Indeed, the community power debate could be criticized as "extremely crude" because of the limitations (Dunleavy, 1980, p. 28).

2.2.2 *Growth Machine*

The growth machine thesis emphasized the role of individuals and interest groups by focusing on the political economy of "place". It outperformed the community power debate by focusing on a broad field not just on what affected the decision maker's decisions. Logan and Molotch (1987) strived to explain aspects of urban change by examining the actions of, and interrelationships between, the main human agents that help produce them. They charged against the community power debate by arguing that human actions were immaterial to social change. Borrowing from classical Marxism, they distinguished use value from exchange value with regard to real properties. Most people recognized the daily use value of their land or building, but a small group of owners went beyond daily use and attempted to gain financial benefit from their assets. These "rentiers" constantly strove to maximize the exchange value of their holdings for higher rents. Because they could not achieve the aims alone, the rentiers create the sort of business climate to attract investment and formed a "growth coalition" with other interested players. What united these players was a commitment to economic growth, based on the tangible benefits to be generated subsequently. Power, in this thesis, refers to the domination over the physical material, intellectual resources and capacity to attract investment. Rentiers who are expected to benefit the most from the project are generally equipped with the strongest power, while other coalition partners deploy power under the former's influence.

The division of users and rentiers and the further categorization of rentiers strengthened the analytical power of the growth machine theory on the actions and interrelationships among agents in urban development. The focus on urban development issues in the growth machine was more specific than in the more diffuse community power studies. With less concern about "who takes control?", the power relations of the theory were more concerned with who had the greatest influence over the physical restructuring of places, why, and with what effect, which made the question of power more manageable.

Whereas several concerns occur when reviewing the growth machine theory. Whether local governments are members of the growth coalition is not clear in this theory though they are active supporters out of concern with increasing growth (Judge et al., 1995). Despite its considerable strengths, the core conceptual assumptions found in the growth machine theory were often criticized for being too simplistic (Clarke, 1990), too elitist and lack of the adoption of multi-participatory decision-making processes and mechanisms (Dowding, 2001; Wood, 2005). Its scope might be in some ways expanded to include the broad politics of development, not just the local government politics of development.

2.2.3 Urban Regime

From a hyper-pluralist perspective, Stone's urban regime thesis conceptualized the coalition of agents as a "regime" (Stone, 1989, 1998). Just as Stone (1986) concluded, complexity was central to the regime, in which diversity, interdependence, fragmentation and lack of consensus characterized the urban system. These characteristics limited the capacity of a state as an agency of authority or control so the government needed to work with other institutions and interests and "[brought] about enough cooperation among disparate community elements to get things done" (Stone, 1989, p. 227). In response to social change and conflict, governmental and non-governmental actors were encouraged to form regimes to facilitate actions and empower themselves. Thus Stone (1989, p. 4) defined regime as "an informal yet relatively stable group with access to institutional resources that enable it to have a sustained role in making governing decisions". As Stone (1986, p. 91) commented, for actors to be effective regime partners two characteristics seemed especially appropriate: first, possession of strategic knowledge of social transactions and a capacity to act on the basis of that knowledge; and second, control of resources that would make one an attractive coalition partner.

In light of urban politics, regime theory produced a new perspective on the understanding of power. In the community power debate, power was about the capacities to blend certain interests to achieve common purposes. The power in the regime was "not control and resistance but gaining and fusing a capacity to act—power to, not power over" (Stone, 1989, p. 229). Stone (1989) conceptualized power as capabilities emerged from effective long-term coalitions to accomplish public aims. "It [directed] attention away from a narrow focus on power as an issue of social control towards an understanding of power expressed through social production" (Judge et al., 1995, p. 55).

Originated from the urban regime theory, Stoker (1998) further put forward four types of power in regime operation: systemic power, command power, coalition power and pre-emptive power[2]. This standpoint extended the implication of power from the perspective of power origin: power was not merely the effect someone took on others as discussed in the community power debate; it deployed resources in the cooperation and coordination between governmental and non-governmental actors in meeting economic and social challenges. However, as coalitions also belonged to a kind of social production, the boundary between coalition power and pre-emptive power was not clear enough and reduced its utility to assume a useful instrument to interpret power with regard to resources.

The framework and the layered conception of power in urban regime theory provided the basis for a subtler understanding of urban politics than the community power debate and the growth machine theory. Power relations referred to the long-lasting and stable relationships to fulfil a public agenda in urban development. However, urban regime theory was often criticized for being excessively localist, absent of the consideration of the rescale of governmental and political relationships caused by globalization and outside issues (Brenner, 1999, 2004). The lack of examination of the consequences of political or corporate embeddedness in multilevel hierarchies was often widely challenged (Pierre, 2014). Furthermore, because of the exaggeration on agency's role and the neglection of institutional structures, regime theory had limited comparative utility in explaining institutional changes and their impact on local actors' interactions (Kantor et al., 1997).

2.3 Conceptualization of Power, Power Resources and Power Relations

The above debates of power relations considered urban development as a social and political vacuum that outside compositions rarely functioned. Two fundamental power theories, namely the structuration theory of Giddens (1984) and the power approach of Foucault (1982), are discussed in this section. They are useful to make connections between broader context and local interactions. Giddens' structuration theory located the power relations analysis into a structure-agent relationship discussion and related localized power interactions to the broader institutional context. Accordingly, the theory provided an informative way to elucidate power and power

---

[2] Systemic power stemmed from the position of actors in the socio-economic structure while command power sprang from the active mobilization of material (information, finance, reputation, knowledge) to dominate over others. Coalition power derived from seeking others who shared compatible goals and complementary resources to bargain in the operation, which always belonged to the less powerful actors. Pre-emptive power originated from the social production, which referred to the capacity to provide leadership and a mode of operation that enabled significant tasks to be done.

resources, but it failed to interpret power relations in a dynamic and localized interacting logic. Foucault considered power deployment as a pervasive relationship that emerged in the operation of deeply rooted dynamic mechanisms. His approach advanced the conceptualization of power relations as a bottom-up process and nested in routine practices. These two theories contributed to the conceptualization of power, power resources and power relations, however, their highly abstractive nature created massive barriers for empirical utility.

2.3.1 Power and Power Resources in the Structuration Theory

The structuration theory interpreted power in the structure-agent debate in which power relations were produced and reproduced from the reciprocity between institutional conditions and the internal power dynamics. Structuralists argued that social structures dominant actors to choose among alternative actions (e.g. Lévi-Strauss & Layton, 1963). On the contrary, some theories overemphasized the freedom of agents to choose their actions despite the limitation of the structured situation (e.g. Marxism). This research agreed with Giddens' position that structure and agents were being shaped by each other—social structure was operating as constraints to actors' action and action (by means of power) worked as the main mechanism constituting and reproducing structures (Giddens, 1979a, 1984; Sewell, 1992).

Giddens signified that power was a transformative individual capability to "act otherwise"—being able to make a difference to interfere or withhold the interference, in terms of affecting "a specific process or state of affairs" (Giddens, 1984, p. 14). Agents with the capacity to act were able to exercise power to influence other actors while the interaction between different agents could react to the structure. The structure was working as constraints on the free initiative of the independent agents (Giddens, 1984).

Power can be understood in a dual-form which includes participants' utilization to succeed (agent power) and the domination in institutionalized structures (structural power) (Giddens, 1979a, 1979b). Giddens attempted to explain how power was generated from the perspective that any assertion of power drew upon two forms of elements: the allocative resources and the authoritative resources (Giddens, 1984). Allocative resources referred to the domination over material facilities such as material features (raw materials), ways of material (re)production (technologies, instruments) and goods production through material features and means. Authoritative resources consisted of transformative capacity generating command over persons or actors (Giddens, 1984). However, there was a huge inconsistency between the dual-form of power and power resources. The original conceptualization of authoritative resources was criticized as being of great abstraction. Although Giddens' followers further explained that authoritative resources included the intellectual with a personal knowledge base, physical strength or social sophistication (Cohen, 1989; Phipps, 2000), the structural constraints of the institution or the formation process of structures were lacking in both resources.

This research appreciates Giddens' contribution to the understanding of power in the interrelation between structure and agents. From this perspective, Giddens' argument was consistent with the opinions of Dahl (1961) and Stone (1989) that power was generated from social interactions, particularly from the mobilization and utility of resources. Giddens also provided a dual way of understanding the power resources which derived from the strategies and approaches participants adopted (i.e. agent power) as well as the institutional structure (i.e. structural power). However, scholars criticized that power was derived from fixed resources rather than stemmed from the daily dynamic interaction in Giddens' theory (e.g. Pred, 1985). Since the discussion of power and power resources remained at a high level of abstraction, it raised the concern that the lack of specificity limited the application of structuration theory in informing empirical research of power relations (Adams & Hastings, 2001; Gregson, 1989; Moos & Dear, 1986; Sewell, 1992). In the next section, Foucault's conceptualization of power and power relations is discussed.

2.3.2 Foucault's Approach to Power

As Foucault (1982) indicated, power had three distinct qualities: its basic nature, its origin and its manifestation (p. 785). The conceptualization of power referred to the basic nature of power, the resources of power applied to power's origin and power relations signified the manifestation of power.

Foucault directed our attention from the preconditions or outcomes of actions to dynamic mechanisms and processes in structured situations. Keeping away from the position that fixed structures assigned definite content to individuals, Foucault (1982) conceived that power was a kind of "capacity" that enabled someone to employ things and gave them the ability to deploy, modify, influence or even destroy. This capacity stemmed from "aptitudes directly inherent in the body or relayed by external instruments" (Foucault, 1982, p, 786). Absent from the sense of structure or agency, Foucault conceived that power was the means whereby all things happen, the production of things, of knowledge and forms of discourse and of pleasure (Cassell, 1993; Foucault, 1982). He challenged the idea that power was exclusively exerted by people or groups who held sovereign domination and regarded power as pervasive (Foucault, 1998) and argued that power was embodied not just in the actors with command over operations but was configured in the people's daily discourses and practices, which in turn might become embedded in institutionalized practices (Bevir, 1999). In this sense, Foucault admitted that the definition of the structural institution could control or regulate the individuals and meanwhile, individuals were able to give effect to a regime of power (structure). This kind of control was a transformative capacity that is consistent with Giddens' conceptualization of power. An essential contribution of Foucault's approach to power was that it transcended politics and saw power as an everyday, socialized and embodied phenomenon. That is to say, power stemmed from a bottom-up process and power relations were expressed in deeply rooted institutionalized interactions between specific actors. Besides, Foucault was hugely influential in pointing that power can originate in different interactions between actors as well as institutional definitions. This was a bottom-up way to conceptualize the power and to interpret the influence of actors' interaction on institutional structure.

Foucault (1982) argued that the exercise of power was not only seen in the relationship between all actors; it also existed when power was put into action to structure the field of other possible actions. His theory guides this research to focus on actors' position, the point of application and the methods used within the political institutions. Therefore, he argued that the analysis of power relations was based on examination of the institution and the emergence and character of an institution could be understood by the contingent, accidental processes whereby various configurations of power keep on changing (i.e. structuring and restructuring themselves). However, without discussing power from the agency-structure perspective, Foucault's approach of power was so elusive. Its character of ubiquity failed to provide tools to study the institutionalization process and there seemed to be of little scope for practical action.

2.4 Institution, Rules and Institutional Analysis

Giddens' structuration theory and Foucault's approach to power both failed to offer concrete devices to inspect the way power being configured by and embedded in the institutions. This section attempts to draw tools from three institutional analysis theories, including the structuration theory of Giddens (1984), the governance approach of Healey et al. (2002) and the Institutional Analysis Development framework of Ostrom (2005), to analyse the institutionalized process in urban development practice.

The focus of institutional analysis is on the devices or policies that give effect to power configuration and the ways in which these devices and policies define, control and regulate the individuals who are subject to the power relations. The structuration theory offered a key tool—social rules -for institutional analysis, however, its ambiguous conceptualization decreased its utility for empirical studies (e.g. Adams & Hastings, 2001). Healey's governance approach contributed to interpreting the localized institutionalization process between the local government and other social actors, but its unidirectional flow and the ignorance of the reaction of social practices to the institution limited its practicality. Ostrom's institutional analysis and development framework provided practical and advantageous tools to scrutinize the repetitive and structured interactions within any institution. However, it ignored the power dynamics and failed to provide enough internal variables, which made it difficult to examine power relations (Rogers, 1989).

2.4.1 Institutional Analysis, Structuration Theory and Rules

Institution can be defined as a social system in which external and local influences coalesced to structure the recurrent changes of actors' interactions in decision-making and social production and reproduction process (Giddens, 1984). The arrangements of an institutional setting refer to the structural forces constructed by social norms, laws, policies, social and physical attributes in any given political, historical and spatial context. They set up a framework to enable the actors and at the same time restrictions to shape their values, the order of their decisions and actions. The institutional analysis assumed the task of identifying the forces in the institutional setting that produced the observed outcomes (Barley & Tolbert, 1997). It focused on the inspection of the relationship between the constraints of the structural properties and their outcomes. Institutionalist approach, therefore, dealt with how actors formed institutions, how institutions operated in practice, and the effects of institutions on the actors over time (Ostrom, 2005).

Institution was traditionally conceptualized as "social system" which granted significant dominations and constraints on the agents (e.g. Talcott, 1951; Von Bertalanffy, 1968). From the perspective of structure-agent debate, Giddens (1979a, 1979b) regarded institutions as "standardised modes of behaviours" (p. 96). These behaviours were working as media and also outcomes of the social practices and subsequently produced the constitution of social systems. He conceptualized the process as "duality of structure", which related to "the fundamentally recursive character of social life" and expressed "the mutual dependence of structure and agency" (Giddens, 1979a, p. 69).

To specify the duality of structure, as Fig. 2.1 indicates, Scott (1995) explained that social systems (or, institutions) not only discharged constraint and empowered the behaviour of actors, but also were reproduced and transformed by the actors' behaviours (p.52). For its constraining ability, institutions created boundaries on the actor's rationality to restrict the probability of their behaviours (Barley & Tolbert, 1997). However, no institution is able to discharge complete control over actors' behaviours as agents possessed capabilities to modify or even eliminate an institution. Agents were enlightened by the restrictions and probabilities which they were confronting in the given context. Their practices and operations, through reactive behaviours, transformed or reproduced the structural properties. The structural configuration and agents' interactions were, therefore, reciprocally influencing each other.

The structuration theory provided two substantial variables—rules and resources—for examining the institutionalization process. Rules are the constraints of the structure. Resources (of the power) refer to the enabling properties of the structure. The continuous production and reproduction of sets of rules and resources through long-time-spans of agents' interactions turned into institutions (Giddens, 1984). Resources serve as the "media" when power is exercised and "as a routine element of the instantiation of conduct in social reproduction" (Giddens, 1984, p. 16). Rules are "the modes of signifying or meaning constitution and normative sanctions"

Fig. 2.1 Duality of structure

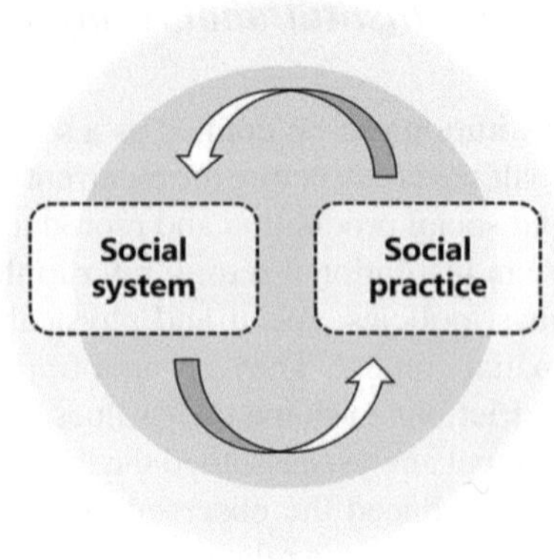

in Giddens theory (Giddens, 1984, p. 28). These rules work as transformational prescriptions to form production and reproduction of social practices, defining "the constitution of meaning" and "the sanctioning of modes of social conduct" (Giddens, 1984, p.18). Compared with the concept of resource, Giddens barely stated a specific conceptualization of institutional rules (Kellerman, 1987; Sewell, 1992). Missing was that the operational conceptualization of rules, that allowed the analysis in what ways rules occurred to influence social practices.

The structure framed regulations and procedures to authorize actors certain power resources to act otherwise. The structure was subject to frequent changes because of other actors' deployment of their own power in reality. The structuring process would form contingent situations as assumed patterns of specific social, spatial or temporal activities. Without sufficient tools, Giddens failed to consider the operation of the institution as a process with a series of interrelated stages and was unable to unveil the power dynamics in specificity (Kellerman, 1987). The ambiguous and vague concept of rules and its lack of specificity reduced its function to describe institution and scrutinize power dynamics in the empirical analysis (Adams & Hastings, 2001; Geoffrey & Hodgson, 2006; Gregson, 1989; Moos & Dear, 1986; Sewell, 1992).

2.4.2 Healey's Institutionalist Framework and Governance

Based on Gidden's structuration theory, Healey created an institutionalist framework to recognize actors, interpersonal relations and organized activities in the governance process (Coaffee & Healey, 2003; Healey, 1992; Healey et al., 2002). Healey conceived the urban governance process as an array of collective complex social interactions shaped by and adjusted to the given historical and spatial context (Healey et al., 2002). In this process, the controlling capacity of the state was diminishing and the role of mobilized public actions at local level was increasing to accomplish the transformation from government to governance (Jessop, 1990). Governance, in other words, was to describe the mechanisms that diffuse configurations of power and authority secured order through the mobilization of collective actors in the absence of state activity (Le Galès, 1998).

Healey's interpretation of "governance" was helpful to understand the interrelationship between the local government and other actors because the traditionally conceived hierarchical and bureaucratical role of the state had been changed to an active and equal one, especially under the influence of market-oriented institutional reform (Healey et al., 2002). The local government diminished its intervention and increasingly relying on other actors to secure policy implementation. The local government and other social actors adopted mobilized practices to get structured in the governance process (Harding et al., 2000). Healey's governance concept provided a new perspective to structure state-society relationship in urban development.

In line with her conceptualization of governance, Healey's institutionalist framework focused on the analysis of the unidirectional impact of institutional change on local policy operation (Healey, 2004, 2006). Without sufficient specific tools, it failed to examine the power dynamic where actors deployed resources and created new rules to alter the structure and was less effective to examine the emerging institutional structure in terms of adaptive behaviour of stakeholders and their changing strategies.

2.4.3 Institutional Analysis and Development Framework

Ostrom proposed the institutional analysis and development (IAD) framework to assist the understanding of nature, the crafted and sustained process and the generated consequences of institutions in diversified context. In this framework, institutions are defined as "enduring regularities of human action in situations structured by prescriptions such as rules, norms, and shared strategies, as well as by the physical world" (Crawford & Ostrom, 1995, p. 582). Ostrom argued that when interdependent actors were behaving in an organized situation, "several layers of components" were created to form the structure and affect their behaviours and the outcomes (Ostrom, 2005, p. 6). In order to easily conceptualize the social structure, the framework dissected the complex adaptive system into nested subassemblies—action arenas, including participants and action situations. Each action arena was composed of nested sets of components such as actors (whether individuals or organizations), actors' actions, an array of decisions and values about process and outcomes. Actors' decisions and interactions impacted the structure of the situations: when outcomes were productive for the involved participants, they tended to maintain the structure of situations. While the participants considered interactions, procedures or outcomes as unfair or inappropriate, they tended to change the structure of the situations (Fehr & Gächter, 2000; Frey et al., 2004). Any action arena would be affected by exogenous factors and generate outcomes which in turn had an impact on the actors and the action situation.

The IAD framework offered a rich variety of variables and tools for scrutinizing the repetitive and structured interactions within any institutions. The framework dissected the complex and long-lasting social structure into subassemblies (arenas) for deeper investigation of the structuration process. Whenever two or more individuals or groups took actions that cooperatively produced or influenced outcomes, these participants could be seen as in an action arena. For an action arena, there are seven clusters of variables: "(1) participants, (2) positions, (3) potential outcomes, (4) the set of allowable actions and the function that maps actions into realized outcomes, (5) the control that participants exercise, (6) types of information generated about actions and outcomes and their linkage, and (7) the costs and benefits assigned to actions and outcomes" (Ostrom, 2005: P, 32). In each situation, all participants have access to some common information, signing that they are in the same situation. Within each characteristic, Ostrom defines their attributes.

Further, by taking the context as external influences and action arenas as internal influences, IAD was able to reveal the reciprocal impacts between the institutional context and local interaction. In her framework, exogenous factors such as the attributes of the physical world, the structure of the community and the institutional arrangement (as the application of rules) acted up in these arenas and affected the structure of arenas (Ostrom, 2005). The attributes of the physical world affected the variables in an action situation in the following ways, for example, what actions and outcomes were possible, what information was useful, how actions were linked to potential outcomes were affected by the context being acted upon in a situation. These factors highlight the impacts of the broader historical and spatial context on local situations, which are not addressed in theories discussed in Sect. 2.1. The attributes of the community were also significant in affecting action arenas including: the sharing values in the community; the common understanding of the structure of action arenas by potential participants; the level of homogeneity of participants' preference in a community; the size and composition of the relevant community; and the scale of inequality of affected participants. The institutional context, as exogenous influences, enabled and constrained social practices in organizational and recurring situations in forms of rules.

Rules, as a substantial variable for institutional analysis, were defined as "prescriptions commonly known and used by a set of participants to order repetitive, interdependent relationships" in the IAD framework (Ostrom, 2005, p. 5). They were accumulated to create the action situation and worked as prescriptions that forbid, permit or require some actions or outcomes and the sanctions authorized if the rules were not allowed (Crawford & Ostrom, 1995; Ostrom, 1986). Ostrom's categorization of rules allowed specifying the concrete mechanism through which actors achieve the capacity to act in this research. In any particular situation, the presence or absence of rules affected the opportunities and constraints individuals confronted, the information they obtained, the benefits they acquired or were excluded from and how they reacted to the situation.

In addition, the evaluation of the governing process in Ostrom's framework provided a reflection of current institutional arrangement which the traditional power relations analysis or other institutionalist theories ignored. In the IAD framework, the institutional analyst might evaluate the outcomes as well as the processes of achieving outcomes. The "evaluative criteria" played a crucial role in the feedback mechanisms. Ostrom indicated that the number of evaluative criteria was large. These evaluative efforts were for individuals to learn from the mistakes and improved the outcomes in the future. The analyst might use different evaluative criteria to conform to different situations, such as equity, efficiency, adaptation and so on.

The IAD framework allowed the researcher to analyse the way actors organize their relationships in the social process and examined relationships between modes of interactions and outcomes in a specific community context, with the influence of exogenous forces. The formation of the action arena was independent of the government's intervention. Therefore, it contained no implicit or explicit biases with respect to how policies were regulated and how institutional constraints were implemented. The different political-economy systems between China and the western world are a barrier to its application, which has been approved by its extensive application to the evaluation of common-pool resources management (Fischer & Qaim, 2012; Ostrom, 1990; Ostrom et al., 1994; Schlager et al., 1994).

The ignorance of duality of structure in the IAD framework, however, decreased its functions in power relations analysis. Just as Rogers (1989) commented, if all power resides ultimately in the properties of inviolate institutional structures, then it was not power at all, since everything was structurally determined without the possibility of effective action on the part of individuals and collectivities. Without conceptualizing the accumulated capacity of the agents—power and its resources, the IAD framework froze its study object by the formation of an action situation under the structuration of the fixed rules as external variables. It treated interactions between actors in the language of rational choice as a stable and fixed gameboard, which different actors interacted to compete for interest (Dowding et al., 2000). In absence of the power dynamic process, it failed to provide enough internal variables to clearly scrutinize how the participants were able to join in the decision-making process and what influence of actors' behaviour granted to the structure. As a consequence, rules in Ostrom's theory only refer to the prescriptions by constitution of meaning, while the Giddensian effects of social interaction was missing.

2.5 Operationalization of the Power Arena Framework

As discussed above, there was not a satisfactory framework to probe the power deployment guiding social changes and the huge inconsistency between broader context and agents' actions. Traditional power relation analysis regarded power dynamics as a linear process of agenda formation, output and impact, ignoring its interaction with the institutional context. Significant power theories such as Giddens' structuration theory and Foucault's power approach failed to provide concrete and specific tools for empirical research for being too abstract. Current institutionalist analytic frameworks, either excessively rely on the analysis of structural constraints or overly emphasize the actors' production and reproduction process. A framework,

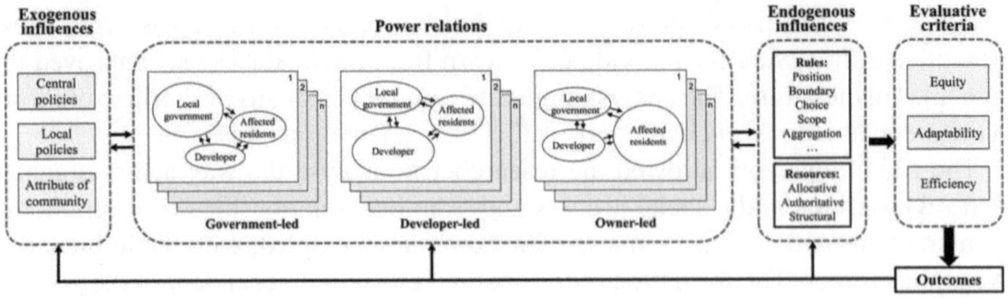

Fig. 2.2 Power arena framework for this book

not only considering the structural enabling and constraining of institutional arrangement, but also analysing the interaction process among the empowered actors and its reaction to the structure is missing.

The analytical framework in this research combines the conceptualization of power, power relations and selectively specific tools in the institutionalist approach. The combination allows the development of a new theoretical lens not only to investigate the enabling and constraining functions of the institutional setting but also to unfold the mobilized participation of the empowered actors in the decision-making process. As the Fig. 2.2 shows, the framework projects the power relations to a dynamic series of cross-sectional "relational arenas". The reciprocal interactions between exogenous and endogenous influences produce and reproduce power relations and generate outcomes, while outcomes of the power interplays will be then evaluated by the principles of equity, institutional adaptability and transaction cost. (Re)development outcomes feed back to the arenas, the exogenous and endogenous influences as an evolving social communicative mechanism. The framework will guide the structure of empirical analysis in Chaps. 4, 5 and 6.

2.5.1 Conceptualization of Power and Power Relations

Power

Two groups of power conceptualization are identified in the review of power-related theories. The first group defines power based on its exercising effect (Bachrach & Baratz, 1962; Logan & Molotch, 1987; Steven, 1974). The second group interprets power as a capacity to act and the capacity relies on the deployment of resources or properties which the actor possesses (Dahl, 1961; Giddens, 1984; Hunter, 1953; Stone, 1998). The conceptualization of power in this research agrees with the idea that power is an acting capacity based on resources deployment. This is because that power (deployed by one actor) manages to influence other actors in various ways with different effects in different contexts and situations. It can be challenging and abrupt to measure the effect and influence of power at the conceptualization stage. This research conceptualizes power based on Giddens (1984)'s structuration theory that power refers to the transformative capacity enabling an actor to act otherwise to achieve their ends through making influences on others in the decision-making process. Power stems from the mobilization of resources the actor possesses. This conceptuality admits the diversity of power and recognizes the adaptability of power to changing contexts. It guides the research of power relations transforms the attention of power from a narrow focus on social control towards a process of social production.

To a great extent, urban development is a profoundly value-related process in which different stakeholders exert power and interact with each other to achieve their own goals. Stakeholders refer to individuals, groups and formal organizations who have a perceived interest or impact on the decision-making. The development process is not individually formed by either government command or the coalition between government and property agency but embodied and displayed in the conflicts, negotiations and compromises among different groups. Stakeholders take collective responsibility for their actions and subsequent outcomes from those actions. The power relations of various participants around the distribution of capital, strategies and interests reflect the nature of the interactions between them, which also determinates the outcome. Based on the structuration theory (Giddens, 1984), power, in this research, is conceptualized as the capacity enabling actor to act otherwise and to achieve their end through the influence on others. It is a transformative capability which is based on the structural configuration or the utilization of resources and enables actors to impose their particular wills in the governing body of the decision-making process. This capacity, therefore, does not exist alone but nests in terms of relationships (Foucault, 1982).

This study adopts inspirations from Giddens' structuration theory and Foucault's power approach in the theoretical construction of power relations. Giddens' structuration theory locates the power relations analysis into a discussion about the structure-agent relationship. It relates the local power interactions among different agents to the formation of external institutional structure (Giddens, 1979a, 1984). The reciprocity between the institutional conditions (structure) and inner power dynamics (agent) produce and reproduce power relations over time (Giddens, 1984). The structure-agent discussion guides the study of power relations to take in consideration both institutional context and actors' interactions. Giddens' theory, however, was criticized as being too structured and neglecting the dynamics of actor's interactions (Pred, 1985). This research utilizes Foucault's power approach to fix the gap. Foucault considers power deployment as a pervasive relationship that emerges in the operation of deeply rooted dynamic mechanisms and his approach conceptualizes power from a bottom-up process. According to Foucault (1982)'s argument, power relations are formed by specific individuals exercise power over other actors who are not inert or powerless. Actors' interactions work as a bottom-up force to proceed with the mutual constitution and transformation of the institution, revealing the reaction, adaptation and reproduction to the structural constraints. Building upon both Giddens and Foucault's theories, this research conceptualizes power relations as a dynamic and continuously relational process with recurrent changes, composed of the continuous mutual reaction between the institutional constraints and agents' interactions.

2.5.2 *Power Relations Analysis*

I. Three Patterns of Power Relations

The participatory reform in China profoundly transforms the mechanisms of urban redevelopment in terms of the roles of the involved actors, their configured power and the interactions they are undertaking. According to current studies, the *San Jiu Gai Zao* policy constrains the power of the local government which is transformed to be a regulator and facilitator. The policy also enhances the power of market sectors (especially the private developer) to play a flexible role to initiate and operate the redevelopment. Besides, the policy established public participation and self-redevelopment mechanisms to empower the community to play an essential role in the decision-making process (X. Li et al., 2019; Tian & Yao, 2018; Ye, 2011). Under the restructured power relations, each of the three key stakeholders can exercise their power to initiate redevelopment projects, which forms three patterns of power relations in participatory urban redevelopment: government-led, developer-led and owner-led redevelopment, as the Fig. 2.3 shows. The review in Sect. 1.2 provides empirical basement to establish the analytical framework to structure the involved actors and their interactions in the participatory redevelopment process.

In the government-led redevelopment, the local government is configured with the dominant power as it takes charge of land expropriation from the existing land user and then auctions it to the potential land user. The power of the residents is strengthened by the participatory policies. Therefore, they are able to adopt several measures to counterbalance the power of the resource-rich actors, e.g. the local government and developers, fighting for fairer compensation terms. The developer is supposed to form a coalition with the local government to deal with administrative complexity and facilitate the accomplishment of the redevelopment projects.

The participatory policy allows the direct land use transfer between the existing land user and the potential land user, which encourages the developer to play an

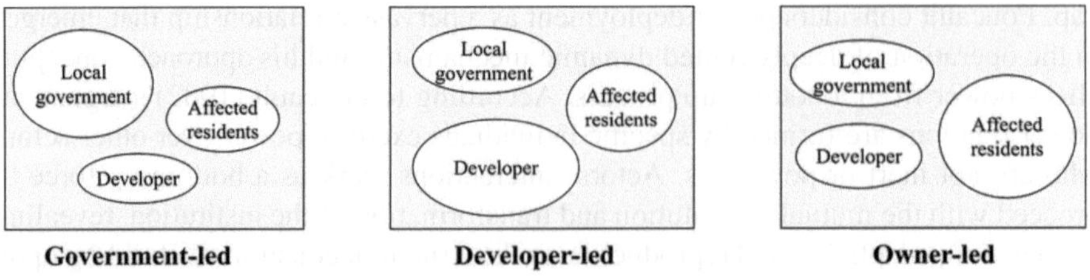

Fig. 2.3 Three patterns of power relations in participatory redevelopment

initiating role to operate the developer-led redevelopment (Ye, 2011). The developer actively negotiates with the affected residents on redevelopment plan details to obtain their agreements. The affected residents are included to join in the decision-making process and bargain with the developer for satisfying compensation terms. In this regard, the government is supposed to play a facilitator role with minimized intervention to promote the property right exchange between the two contracting parties.

In the owner-led redevelopment, residents have been authorized the self-redevelopment right by the participatory reform to initiate owner-led redevelopment project. The increased autonomy enables them to share a larger portion of interests from the land value increment in urban redevelopment (Tian & Yao, 2018). In this respect, residents are supposed to play the role of initiator to collect support from the majority of the affected residents. As the property rights are fragmentedly owned by individual households, the affected residents may form a collective association as a representative to negotiate with the local government and the developer. Three parties tend to form a collaborative partnership to accomplish the redevelopment project. The above three paragraphs indicate the hypothetical primary involved stakeholders and power relations in three different patterns of residential redevelopment, guiding the formation of the analytical framework in this dissertation.

II. Dissection of Power Arenas

To specifically examine the decision-making process, this framework dissects power relations into a dynamic series of cross-sectional "relational arenas". The dissection of power arenas is inspired by Ostrom to unpack the development process into its constituent sections to recognize different accumulations of social relations which might surround each arena. Each arena is considered as a focal unit of the development process, concerning who is empowered to take actions or affect outcomes related to a specific domain. Multiple arenas linked sequentially or simultaneously and exist dependently of each other to compose of the social reality, as the Fig. 2.4 shows.

The common information all participants have access to signify the formation of the arena (Ostrom, 2005). Within each power arena, stakeholders (such as the local government, the developer and affected residents) interact with each other according to their private interests and values. The division of the power arena is subject to

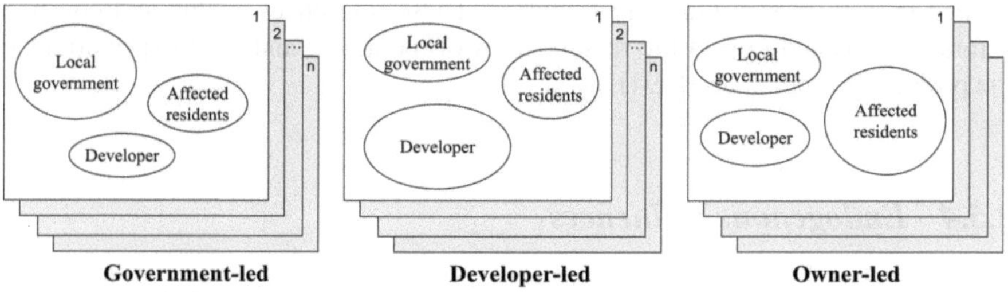

Fig. 2.4 The dissection of power arenas

considerable sectoral, spatial and temporal variation. In this research, the dissections of power arenas are modified to respond to particular situations, with regard to the initiator's strategy or the changing organizers of the arena. When the interactions yielding outcomes are productive for those involved, the actors may increase their commitment to maintaining the structure to continually receive benefits. When actors view interactions as inappropriate or fail to achieve their ends, they may change their strategies and subsequently transform the structure. Taking account of spatial and temporal variation, the dissection of power arenas is able to allow assessing the particular "driving dynamics" in distinctive patterns of agent relations. Owing to the adaptability of arena dissection, this framework is capable of the application under different economic and political contexts.

2.5.3 *Exogenous Influences*

The exogenous influences directly affect operational situations in the power arenas by creating and limiting participants' actions, and by deciding who is presenting power and with what weight in the decision-making process. They reveal a top-down instruction of the interactions in power arenas but normally do not respond swiftly to the change of power dynamics at the local level. The existence of exogenous influences results in the uneven distribution of power on the structural side.

As discussed in the urban power debates, important exogenous factors cannot be ignored in the power relations analysis. This framework classifies exogenous factors into three main themes, central policies, local policies and attribute of the community, to improve the categorization of the exogenous factors, as Fig. 2.5 displays. The central policies bring about great changes in structural properties and governance mechanisms, such as housing reform, the institutionalization of property rights, participatory urban redevelopment reform and the abolition of demolition permit institution. Local policies in the land use and planning field materialize and modify the governance systems at local level. The attributes of the community were significant in affecting action arenas including the physical attributes and social compositions of the community. The social compositions are composed of social status, social bond, social norms and values of the affected residents. They manifest not only material constraints of a particular physical relation, but also are deeply enmeshed and framed in a given socio-historical context. These references enable and constrain actors' actions and impact the relationships of the actors and their expectations of the project (Ostrom et al., 1994).

2.5.4 *Endogenous Influences*

Although the institutional setting discharges restrictions on actors' behaviours, local actors hold a certain degree of autonomy from their subjective initiative to make a

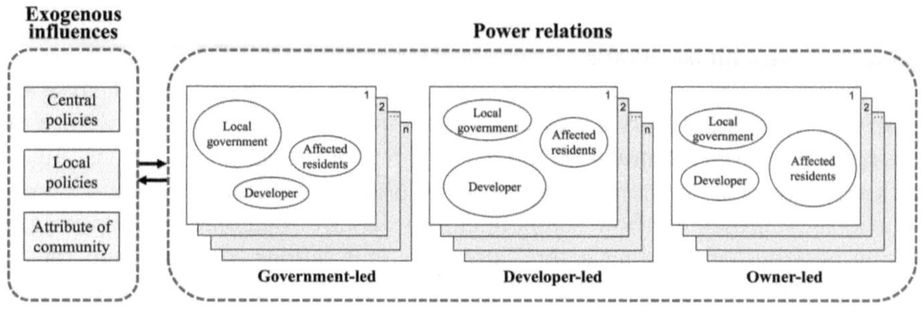

Fig. 2.5 Exogenous influences

difference in the decision-making process. Endogenous influences, which constantly change along with the alteration of actors' power dynamics, affect what decisions they are empowered to make, how they make decisions, what actions they take and what consequence or impact these will lead to. The enabling actors obtain power based on diverse resources and subsequently adopt some strategies and actions, aiming to adapt, react and even change the power relations. Based on Giddens (1984)s structuration theory, two endogenous influences are conceptualized to constitute power relations: power resources and rules. Power resources signify the enabling property from either structural or agent perspective for actors to act otherwise in the interaction. In other words, these possessed resources are the basis to enable actors to deploy their power. Rules signify as channels that the power to take effect through the configuration of strategies and actions. They refer to the manifestations and effects of the adopted strategies and actions to structure and restructure the power relations.

In the urban redevelopment process in China, the origins of different stakeholders' power are diversified. As mentioned in Sect. 1.2, current studies contended that local governments obtain their power from the institutional definition, regulatory authority, administrative jurisdictions, access to land resources, control of financial institutions and a small share of investment in the projects. Developers derive their power from financial capital, expertise, local knowledge, good relationship with the local governments and good reputation. Residents get their power from housing ownership, specific policies, the recruitment in local government. This descriptive way of power origin creates a significant barrier to the comparison of different power resources.

According to the structuration theory, this research categorized power resources into three types: structural resources, allocative resources and authoritative resources. The structural power derives from the definition of structural properties such as institution and policy definition while the agent power is built upon allocative and authoritative resources (Giddens, 1984). Allocative resources refer to the domination over material facilities such as financial strength and expertise. Authoritative resources consist of the human command over others, including constitution of chances of self-development and self-expression such as knowledge, skills and experts and organizations and relations of human beings in mutual association such as belief, social norms, reputation, authority, social relationships and networks (Cohen, 1989; Giddens, 1984; Phipps, 2000).

Since current theories fail to provide a useful instrument to interpret, analyse and compare actors' behaviours in power relations studies, in this research, we use the concept of rules to conceptualize the actor's empirical strategies and actions. Rules[3] work as approaches by the initiative actors to order or structure their counterpart's behaviours, representing how actors attempt to control outcomes. By opening or closing possibilities of other actors' strategies and actions, rules account for the exercise of power to unfold in the structured situations. Any action necessarily implies an incentive, reaction, stimulation or resistance being a part of the reproduction process, thus revealing a type of rule. These rules, working as means to change the structure of incentives, express the effects of agents' efforts to impose their wills in standardized situations. The conceptualization of rules enables us to interpret and contrast the meaning and functions of the fragmented strategies and actions nested in the power configuration.

The categorization of rules by Ostrom proposes an approach to precisely structuring how power functions. In varying forms, rules are applied to an array of actions, such as the use of (1) position rules to specify positions for participants in development process, (2) boundary rules to define how participants are chosen to or leave these positions, (3) choice rules to constrain other actor's actions, (4) scope rules to allow or forbid the effect of other actor's actions, (5) aggregation rules to unite actors towards a collective action, (6) information rules to define approaches of communication among actors and (7) pay-off rules for the distribution of interests and costs among participants (Ostrom, 1986, 2005). The conceptualization of power into rules enables the discussion of frequency and intensity of power and facilitates the cross-case comparison.

The configuration of rules and resources results in a bottom-up force to influence the structural properties, as the Fig. 2.6 exhibits. By disclosing the power relations in a social process, this framework are able to examine what mechanisms regulate the discharge of power and what are the impacts of power relations on the outcome.

2.5.5 *Outcomes*

Power relations are working as the production of the interplay between structural properties and social practices, and also feedback to actors' interactions and institutional structure. The outcome of the previous arena may create new rules or resources for the practice in the next arena. Meanwhile, the recursive relation in the next power

[3] Rules are also able to be applied to the interaction between exogenous and endogenous influences, or how endogenous actors respond to exogenous changes. In this research, the focus is placed on the power relations at the micro-level of urban redevelopment. Therefore, rules are restricted to endogenous influences about actions and behaviours of involved actors.

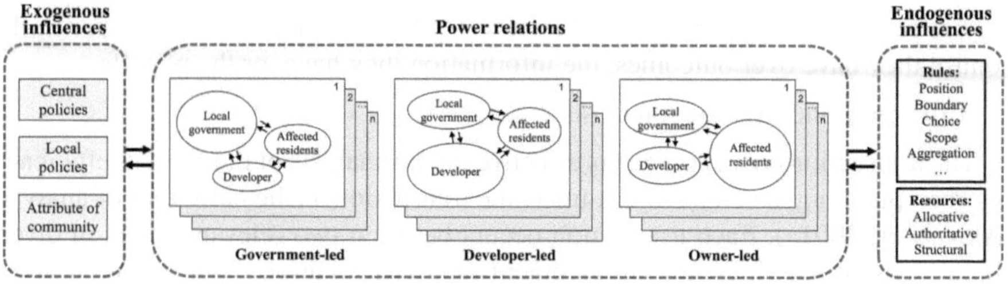

Fig. 2.6 Mutual functions between exogenous influences and endogenous influences

arena among stable sets of rules and social practices is the result of the operation of power relationships in the previous one.

To improve the unclear conception of "outcome" in Ostrom's framework, this research conceptualizes outcome as spatial and social outcomes. It spatially includes layout, density (Floor Area Ratio), function and living environment of the community and also comprises of the social impacts of the spatial outcome.

I. Evaluative Criteria

Good institution and governance require accountability, which necessitates effective monitoring and evaluation. The power arena framework, in this research, is also aiming to function as an analytical framework to scrutinize the structure of institutional arrangement and its performance and outcome. To monitor and evaluate the residential redevelopment mechanisms in Shenzhen, this framework chooses three evaluative criteria as follows: equity, adaptability and efficiency.

a) Equity

The assessing of equity is on the basis of the equality between individuals' contributions to an effort and the benefits they garner with the principle that those who derive greater benefits are expected to pay more (Imperial, 1999b). This criterion is to measure the redistribution effects of interests generating from the redevelopment among stakeholders.

b) Institutional Adaptation

Through the factor adaptability, this framework measures stakeholders' adaptation to address path dependence. The interaction among stakeholders in each arena is highly competitive, while stakeholders deploy power to achieve their ends. When actors perceive interactions as inappropriate or unfair, they may change their strategies for a positive outcome. Actors who adopted better rules will enhance their benefit while those who do not search for alternative rules will lose out (Ostrom, 1990). If substantial exogenous changes occur and change the structure of the arena, actors will be reselected and further restructure the power configuration. As a consequence, individuals learn from experiences in one arena and adapt to new circumstances by developing new strategies leading towards possible norms and rules, thus modify power structure and resource distribution (Ostrom et al., 1993). The changing rule may affect the set of participants, the set of strategies available to participants, the control they have over outcomes, the information they have, or the payoffs.

c) Efficiency and Transaction Costs

The transaction cost approach is effective on the analysis of the process efficiency of urban planning development. It treats the transaction as the basic unit of analysis (Alexander, 2001a). Each transaction occurs between two relevant actors in diversified types of activities: the procurement or transformation of information about the values of resources, the goals and demands of the counterpart; the procurement of services, ideas and intellectual capitals; the fundamental negotiation to exchange information; the transfer or distribution goods and people (Alexander, 2001b; Coase, 1960).

This research measures the process efficiency and transaction costs from the aspect of actors' behaviours and power relations. The degree or level of transaction costs is affected by power distribution, opportunism and heterogeneous interests. Power symmetry is understood as a balance between power-holders taking into consideration the strength of the stakeholder measured by number (e.g. the number of households), the resources that they command (e.g. consent to sign an agreement) and the position (e.g. dominant) that they have in the relationship. When power asymmetry emerges or where the rules and sanctions are lacking, opportunistic behaviours occur in trading activities (Imperial, 1999a). In this regard, the asymmetries of power refer to the situation that one side exercises an overwhelming power to exclude or force the other side (Imperial, 1999b).

Opportunism is the self-interestedness of economic actors who take interest-seeking behaviours with deceit and absence of moral restraints (Williamson, 1993). It involves deliberately withholding or distorting important business information, shirking, e.g. doing less work than agreed, or failing to fulfil formal or informal promises and obligations. It raised the supervision cost of the transaction process and low exchange efficiency. Heterogenous interests create barriers to reaching group consensus, which is influenced by size of group, numbers of stakeholders, institutional context and power distribution among parties, etc. (Alexander, 2001b; Buitelaar, 2004).

References

Adams, D., & Hastings, E. (2001). Assessing institutional relations in development partnerships: The land development corporation and the Hong Kong government prior to 1997. *Urban Studies, 38*(9), 1473–1492.

Alexander, E. R. (2001a). Governance and transaction costs in planning systems: A conceptual framework for institutional analysis of land-use planning and development control—the case of Israel. *Environment and Planning b: Planning and Design, 28*(5), 755–776.

Alexander, E. R. (2001b). A transaction-cost theory of land use planning and development control: Towards the institutional analysis of public planning. *Town Planning Review, 72*(1), 45–76.

Bachrach, P., & Baratz, M. S. (1962). Two faces of power. *American Political Science Review, 56*(04), 947–952.

Bachrach, P., & Baratz, M. S. (1970). *Power and poverty: Theory and practice*. Oxford University Press.
Barley, S. R., & Tolbert, P. S. (1997). Institutionalization and structuration: Studying the links between action and institution. *Organization Studies, 18*(1), 93–117.
Bevir, M. (1999). Foucault, power, and institutions. *Political Studies, 47*(2), 345–359.
Blokland, T. (2003). *Urban bonds*. Social relations in an inner city neighbourhood.
Brenner, N. (1999). Globalisation as reterritorialisation: The re-scaling of urban governance in the European union. *Urban Studies, 36*(3), 431–451. https://doi.org/10.1080/0042098993466
Brenner, N. (2004). Urban governance and the production of new state spaces in Western Europe, 1960–2000. *Review of International Political Economy, 11*(3), 447–488.
Buitelaar, E. (2004). A transaction-cost analysis of the land development process. *Urban Studies, 41*(13), 2539–2553.
Cassell, P. (1993). Domination and power. In *The giddens reader* (pp. 212–283). Springer.
Clarke, S. E. (1990). 'Precious'place: The local growth machine in an era of global restructuring. *Urban Geography, 11*(2), 185–193.
Coaffee, J., & Healey, P. (2003). My voice: My place': Tracking transformations in urban governance. *Urban Studies, 40*(10), 1979–1999. https://doi.org/10.1080/0042098032000116077
Coase, R. H. (1960). The problem of social cost. *The Journal of Law and Economics, 56*(4), 837–877.
Cohen, I. J. (1989). *Structuration theory: Anthony Giddens and the constitution of social life*. Macmillan London.
Crawford, S. E., & Ostrom, E. (1995). A grammar of institutions. *American Political Science Review, 89*(3), 582–600.
Dahl, R. A. (1961). *Who governs?: Democracy and power in an American city*. Yale University Press.
Dowding, K. (2001). Explaining urban regimes. *International Journal of Urban and Regional Research, 25*(1), 7–+. https://doi.org/10.1111/1468-2427.00294
Dowding, K., Dunleavy, P., & King, D. (2000). Understanding urban governance: The contribution of rational choice. In G. Stoker (Ed.), *The new politics of British local governance* (pp. 91–116). Palgrave Macmillan.
Dunleavy, P. (1980). *Urban political analysis: The politics of collective consumption*. Springer.
Fehr, E., & Gächter, S. (2000). Fairness and retaliation: The economics of reciprocity. *The Journal of Economic Perspectives, 14*(3), 159–181.
Fischer, E., & Qaim, M. (2012). Linking smallholders to markets: Determinants and impacts of farmer collective action in Kenya. *World Development, 40*(6), 1255–1268. https://doi.org/10.1016/j.worlddev.2011.11.018
Foucault, M. (1982). The subject and power. *Critical inquiry, 8*(4), 777–795
Foucault, M. (1998). The will to knowledge. *The History of sexuality*, vol. 1, trans. Robert Hurley. Penguin.
Frey, B. S., Benz, M., & Stutzer, A. (2004). Introducing procedural utility: Not only what, but also how matters. *Journal of Institutional and Theoretical Economics JITE, 160*(3), 377–401.
Geoffrey, M. H., & Hodgson, R. (2006). What are institutions? *Journal of Economic Issues, 40*(1), 1–25.
Giddens, A. (1979a). *Central problems in social theory: Action, structure, and contradiction in social analysis* (Vol. 241). Univ of California Press.
Giddens, A. (1979b). *Studies in social and political theory* [Bibliographies Non-fiction]. Hutchinson.
Giddens, A. (1984). *The constitution of society: Outline of the theory of structuration*. Univ of California Press.
Gregson, N. (1989). On the (ir) relevance of structuration theory to empirical research. In *Social theory of modern societies: Anthony Giddens and his critics* (pp. 235–248).
Harding, A. (1995). *Elite theory and growth machines*. Sage.

Harding, A., Wilks-Heeg, S., & Hutchins, M. (2000). Business, government and the business of urban governance. *Urban Studies, 37*(5–6), 975–994.
Haugaard, M. (2002). *Power: A reader*. Manchester University Press.
Healey, P. (1992). An institutional model of the development process. *Journal of Property Research, 9*(1), 33–44.
Healey, P., Cars, G., Madanipour, A., & De Magalhaes, C. (2002). Transforming governance, institutionalist analysis and institutional capacity. In *Urban governance, institutional capacity and social Milieux* (pp. 20–42). Routledge
Healey, P. (2004). Creativity and urban governance. *disP-The Planning Review, 40*(158), 11–20.
Healey, P. (2006). Transforming governance: Challenges of institutional adaptation and a new politics of space. *European Planning Studies, 14*(3), 299–320.
Hunter, F. (1953). *Community power structure A study of decision makers*. The University of North Carolina Press.
Imperial, M. T. (1999a). Analyzing institutional arrangements for ecosystem-based management: Lessons from the Rhode Island Salt Ponds SAM Plan. *Coastal Management, 27*(1), 31–56.
Imperial, M. T. (1999b). Institutional analysis and ecosystem-based management: The institutional analysis and development framework. *Environmental Management, 24*(4), 449–465.
Jessop, B. (1990). *State theory: Putting the capitalist state in its place*. Penn State Press.
Judge, D., Stoker, G., & Wolman, H. (1995). *Theories of urban politics*. Sage.
Kantor, P., Savitch, H. V., & Haddock, S. V. (1997). The political economy of urban regimes: A comparative perspective. *Urban Affairs Review, 32*(3), 348–377.
Kellerman, A. (1987). Structuration theory and attempts at integration in human Geography∗. *The Professional Geographer, 39*(3), 267–274.
Le Galès, P. (1998). Regulations and governance in European cities. *International Journal of Urban and Regional Research, 22*(3), 482–506.
Lévi-Strauss, C., & Layton, M. (1963). *Structural anthropology* (vol. 1). Basic books New York.
Li, X., Han, S. S., & Wu, H. (2019). Urban consolidation, power relations, and dilapidated residential redevelopment in Mutoulong, Shenzhen, China. *Urban Studies, 56*(13), 2802–2819. https://doi.org/10.1177/0042098018799950
Logan, J. R., & Molotch, H. (1987). *Urban fortunes: The political economy of place*. Univ of California Press.
Moos, A., & Dear, M. J. (1986). Structuration theory in urban analysis: 1. *Theoretical exegesis. Environment and planning A, 18*(2), 231–252.
Ostrom, E. (1986). An agenda for the study of institutions. *Public Choice, 48*(1), 3–25.
Ostrom, E. (1990). *Governing the commons: The evolution of institutions for collective action*. Cambridge University Press.
Ostrom, E. (2005). *Understanding institutional diversity*. Princeton University Press.
Ostrom, E., Gardner, R., & Walker, J. (1994). *Rules, games, and common-pool resources*. University of Michigan Press.
Ostrom, E., Schroeder, L., & Wynne, S. (1993). *Institutional incentives and sustainable development: infrastructure policies in perspective*. Westview Press.
Phipps, A. G. (2000). A structurationist interpretation of community activism during school closures. *Environment and Planning A, 32*(10), 1807–1823.
Pierre, J. (2014). Can urban regimes travel in time and space? Urban regime theory, urban governance theory, and comparative urban politics. *Urban Affairs Review, 50*(6), 864–889. https://doi.org/10.1177/1078087413518175
Polsby, N. W. (1980). *Community power and political theory*.
Pred, A. (1985). The social becomes the spatial, the spatial becomes the social: enclosures, social change and the becoming of places in Skåne. In *Social relations and spatial structures* (pp. 337–365). Springer.
Rogers, A. (1989). *Gentrification, power and versions of community: A case study of Los Angeles*. University of Oxford.

Schlager, E., Blomquist, W., & Tang, S. Y. (1994). Mobile flows, storage, and self-organized institutions for governing common-pool resources. *Land Economics*, 294–317.
Scott, W. R. (1995). *Institutions and organizations. Foundations for organizational science*. The Sage
Sewell, W. H., Jr. (1992). A theory of structure: Duality, agency, and transformation. *American Journal of Sociology, 98*(1), 1–29.
Steven, L. (1974). *Power: A radical view*. Macmillan.
Stoker, G. (1998). Theory and urban politics. *International political science review/Revue internationale de science politique*, 119–129.
Stone, C. N. (1986). Power and social complexity. In *Community power: Directions for future research* (pp. 77–113).
Stone, C. N. (1989). *Regime politics: Governing Atlanta, 1946–1988*. Univ Pr of Kansas.
Stone, C. N. (1998). Regime analysis and the study of urban politics, a rejoinder. *Journal of Urban Affairs, 20*(3), 249–260.
Talcott, P. (1951). *The social system*. Routledge.
Tian, L., & Yao, Z. (2018). From state-dominant to bottom-up redevelopment: Can institutional change facilitate urban and rural redevelopment in China. *Cities, 76*, 72–83.
Von Bertalanffy, L. (1968). General system theory. *New York, 41973*(1968), 40.
Waste, R. J. (1987). *Power and pluralism in American cities: Researching the urban laboratory* (Vol. 165). Greenwood Publishing Group.
Weber, M. (1947). *The theory of social and economic organization*. Simon and Schuster.
Wellman, B., & Leighton, B. (1979). Networks, neighborhoods, and communities: Approaches to the study of the community question. *Urban Affairs Quarterly, 14*(3), 363–390.
Williamson, O. E. (1993). Opportunism and its critics. *Managerial and Decision Economics, 14*(2), 97–107.
Wood, A. (2005). Comparative urban politics and the question of scale. *Space and Polity, 9*(3), 201–215.
Ye, L. (2011). Urban regeneration in China: Policy, development, and issues. *Local Economy, 26*(5), 337–347. https://doi.org/10.1177/0269094211409117

Chapter 3
Urban Redevelopment Policies in Shenzhen, China

3.1 Introduction

The opening-up and market-oriented reforms of the 1980s decentralized authority to local governments, enabling municipal authorities to assume a dominant role in urban development. Local governments subsequently pursued a dual-track urbanization strategy to facilitate rapid urban growth, comprising inner-city redevelopment within the established urban core and extensive suburban expansion in peripheral areas. This accelerated urbanization generated substantial fiscal revenues for local governments and contributed significantly to local economic development. However, it also produced a series of negative consequences, including inefficient land use and environmental pressures. In response, governments at various administrative levels introduced policies aimed at curbing low-density urban sprawl on non-urbanized land while promoting higher-density development within the existing urbanized areas. Together, these interventions signalled the onset of a new era of urban redevelopment in China.

This chapter introduces the institutional context of urban redevelopment. Section 3.2 presents the background, processes and consequences of rapid urban development. Section 3.3 examines the efforts of the central and provincial governments in formulating urban redevelopment policies. Section 3.4 analyses the measures adopted by the Shenzhen municipal government to facilitate the implementation of participatory urban redevelopment. Section 3.5 explores the various mechanisms of dilapidated residential redevelopment in Shenzhen. Finally, Sect. 3.6 concludes the chapter.

X. Li, *Participatory Redevelopment as Contested Arenas in China*,
https://doi.org/10.1007/978-981-95-9394-1_3

3.2 Rapid Urban Development Following Reform and Opening-up in China

The adoption of the land leasing system and housing commodification, alongside the decentralization of administrative power in the late 1980s encouraged the rapid urban development process in two dimensions: urban redevelopment in inner areas and urban expansion in periphery areas. This section analyses the processes and consequences of this accelerated urbanization.

3.2.1 Institutional Context of China's Reform and Opening-up

In the late 1970s, the Chinese Central State launched a marketization and opening-up reform to salvage the failing society under the centrally controlled system riddled with malinvestment and inefficiencies (Brandt & Rawski, 2008). The reform released the constraints on the market power and restructured the obligation arrangement between the central and local states, which discharged substantial influences on urban development (Yeh & Wu, 1996).

The impacts on urban development were commenced in 1987 when the central government established the land leasehold market (Yeh & Wu, 1996). Market value has been introduced to the land in urban area and the transfer of land use right was no longer free of charge. The introduction of land leasing system was to rationalize the allocation of land and improve land-use efficiency. However, the presence of price differentials between the administrative land allocation and commercial land transfer generated prevalent speculative behaviours: some public institutes obtained allocated land and sold it in the land market (Yeh & Wu, 1996; Zhang, 2002). In 1991, the central policy institutionalized the government's monopoly in the primary land market—any potential land user should purchase land use right from local governments and the land transfer between other users was forbidden (Yeh & Wu, 1996). Land use transfer thus has turned into an important means of capital accumulation for the local governments. The latter embarked on a massive land leasing push, which triggered the rapid urbanization process.

The commercialization and privatization of the public housing provision also facilitated rapid urban development. The year 1988 marked the beginning of housing reform when the State Council promulgated a policy to initiate the commercialization of housing (He & Wu, 2009). In the 1990s, a series of reform policies launched a housing provident fund system to increase employees' wage for housing purpose and the welfare housing system was gradually eliminated (Wang, 2001; Wang & Murie, 1996). In 1998, the housing monetarization policy was established by the central government to encourage monetary distribution and abolish the in-kind housing subsidy (Wang, 2001). Since then, the responsibility of housing provision transferred to the housing market while the government and work units were no longer required to meet the housing requirements of their employees. In 1999, the central state published a milestone document to promote the privatization of public housing (Wang et al., 2012). It encouraged the local housing authority to sell public housing to the sitting tenant. As a response to the transferrable land and commercialized housing, a real estate market has been created and soon becomes one of the major growth poles in urban development (Ding & Knaap, 2005; Shin, 2009).

To shift development pressure and increase efficiency, the central state decentralized the administrative power to form new incentives at local level. Through the enactment of the Urban Planning Act in 1989, the central government authorized the power of urban planning establishment and land use permit issuance to the municipal government (He & Wu, 2009). In 1994, a tax-sharing system was promulgated to redistribute the revenue between the central state and the local government (Oi, 1992). According to the new system, the central government withdrew a large portion of tax-related revenue while the share of local government was decreased (Wong, 2000). Meanwhile, to create incentives at local level, the tax-sharing system enhanced the financial autonomy of the local government and arranged all land-based taxes[1] to the latter. Reducing financial support from the superior and increasing autonomy incentivized the local government to play an entrepreneurial role to be responsible for their own profits and losses (Harvey, 2002). The local government commenced to substantially promote the land-related development at local level; land leasing associated with space production thus has been the principal approach to amplifying the revenue of the municipalities (Hsing, 2010). Vertical forces of decentralization and marketization from the central level triggered intensive horizontal competitions between different local governments (He & Wu, 2009; Zhou & Logan, 2008; Zhu, 2004). Land-related campaigns were initiated from the urban core to the urban fringe for continuous capital accumulation.

3.2.2 Enforced Demolition and Social Instability in Urban Redevelopment

The land reform marketized the value of urban land and encouraged Chinese municipal governments to initiate large-scale urban redevelopment projects in the built-up area. Through these projects, local states not only aimed to activate the commercial value of the inner city area but also to restructure the urban functions for luring greater resources in the local competition (Hsing, 2010; Wu, 2004).

In the 1980s and the 1990s, with a severe resource to improve the housing condition, local governments began to introduce market sectors to redevelop the deteriorated housing in inner city (Dowall, 1993, 1994). Because of the presence of public housing system, the local government tried to avoid the emergence of homeless in

[1] Land-based resources are including tax on urban land use and occupancy of rural land, increment tax on land value, housing property tax, urban construction and maintenance tax and so on (Wong, 2000).

urban redevelopment projects. Therefore, urban redevelopment policies particularly emphasized on in-kind and on-site compensation (Dowall, 1994). The local government had the decision-making power to acquire land and delegate developers to demolish, construct and compensate (Yeh & Wu, 1996). Developers, as state-owned companies or joint ventures, were requested and often supervised by the government to undertake the redevelopment projects. They were required to provide numerous taxes and fees, as well as in-situ resettlement and improved public facilities free of charge to the local government. The affected tenants, despite the absence of property rights, were able to relocate to better quality housing and continued to pay low rents (Dowall, 1994). This incentive made the relocation process generally smooth without severe conflicts (Wu, 2004). The mechanism heavily biassed towards sitting tenants and municipal government but impaired developers' project financial feasibility (Dowall, 1993, 1994). Although local governments provided an array of incentives, such as higher plot ratio allowance and internal transfer, the "missing profit" demotivated developers, which resulted in supply shortage and severe project delay.

In order to deal with the developer's incentive problem in residential redevelopment, a set of policies was promulgated to initiate efficient large-scale urban redevelopment. Through the housing reform in the late 1990s, the central state abolished the public housing system and transferred the housing provision responsibility to the real estate industry (Wang & Murie, 1996). The original tenants purchased the housing units from their employers to become de facto property owners. In 2001, a demolition permit scheme was established to retain the local government's authority for compulsory demolition and move from resettlement provision to monetary compensation (Fang & Zhang, 2003). Through the adoption of monetary compensation, the policy aimed to increase the feasibility of urban redevelopment, accelerate the redevelopment process and provide more options for the affected residents. Nevertheless, in practice, the policy fostered the interest-seeking behaviours of the economically powerful stakeholders (local states and developers) and impaired the interests of the affected residents. The developers' aim of project acceleration for cost reduction colluded with the local government's objective of rapid capital accumulation and economic growth. Both parties tended to form a growth coalition to facilitate urban redevelopment (Fang & Zhang, 2003). The local government firstly determined areas that were designated to be renovated without any acquaintance of the affected property owners (Wu, 2004). Through the issuance of demolition permit, the municipal government reduced the complexity of different jurisdiction applications and delegated the authorities of compulsory demolition to the developer (Han & Wang, 2003; Zhang, 2008). Without bargaining power, the original residents were obligated to enter a contract and comply with the administrative decision for demolition and relocation (Yang & Chang, 2007). Once the official notice for residents' demolition and relocation was issued on, demolition and displacement always took place within a short period (Shin, 2009).

Under the influence of the policy, urban redevelopment thus became a means of growth promotion to stimulate rapid land-centred capital accumulation and extraction for the economically powerful players (He & Wu, 2009). The state shifted its role from a social welfare provider to an interventionist to regulate market operation, while the obligation of urban redevelopment changed from the municipal governments or work units to private sectors. The lack of public participation in urban redevelopment resulted in unbalanced interest distribution. Through space creation and capital circulation, developers gained immense revenues from the sale of commodity housing while the governments obtained improved public infrastructure free of charge and gained considerable profit through land use fee and relevant taxes (Zhou et al., 2018). Residents were either provided with in-situ housing in the urban fringe or monetary reimbursement which was far below the price of the newly constructed on-site apartments (Yang & Chang, 2007). The low level of public participation in the decision-making process and unfair interest distribution triggered intensive social resistance, which resulted in violent conflicts and severe social instability (Ye, 2011).

To resolve the violent conflicts in the rapid redevelopment, from late 2003, an array of central ordinances were promulgated to forbid the enforced relocation, protect the right-to-know and appropriate compensation of the affected residents (He & Wu, 2009). However, these efforts failed to solve the social instability due to the persistence of the local government's entrepreneurial essence which was facilitated by the government-monopolized land transfer and top-down land use planning mechanism (Shin, 2009).

The aggressive promotion of urban (re)development at local level also caused tensions between the central and local governments (He & Wu, 2009). At the central level, the market-oriented reform was a trial-and-error experiment to decentralize the development pressure and improve land-use efficiency. However, the local government attempted to seek short-term returns and extract value from the urban redevelopment process. Speculative and corrupt practices were not rare of ordinary occurrence (He & Wu, 2009). The issues and contradictions of the efficiency-oriented redevelopment needed to address.

3.2.3 Rapid Urban Expansion and Consequent Issues

The transformation from a centrally planned economy to a market-oriented economy made the local government an entrepreneurial role with an urgent desire of economic growth and revenue accumulation (Wu et al., 2006). The municipalities hence promoted an accelerated urbanization process accompanied by massive rural land conversion to urbanized land (Lin, 2009). The land-related institutional arrangement enhances the urban government control over rural land and thus facilitates rural land conversion. Although the village collective ownership of land in rural area is legitimated and protected by the Constitution, the collective cannot transfer rural land to outside investors due to the constraints of land use policies (Hsing, 2010). The land management law further delegates municipal governments the power to require rural land to lease out and the requisition has to be out of public interests (Hsing, 2010). However, the ambiguous definition of public interests created spaces for the local state to conduct speculative behaviours. The latter commenced expropriating farmland with the endorsement of the "public interest" for commercial development and capital accumulation (Hsing, 2010). In this process, urban planning has been utilized by local governments as a proactive and strategical instrument to facilitate the land conversion.

The Ministry of Land Management allocates rural land conversion quotas to governments at lower level through the establishment of national annual land use plans. By means of the prediction of population and economic growth in the land use plan, the local land authority bargained with governments at higher level for urban construction land quotas. The obtained land quotas provide legalized permissions for the local governments to undertake occupation and conversion of the cultivated agricultural land in the urban fringe, which triggers rapid urban expansion (Han, 2010; Yue et al., 2013; Zhang, 2000). Industrial parks, economic and technological development zones, city-led comprehensive residential development and new towns have evolved into a fully-fledged campaign (Wu et al., 2006).

Continuous and vast rural land conversion created severe urban sprawl in China. Research reveals that the total urban land area increased from 203.66 km^2 in 1980 to 1370.07 km^2 in 2010 (Lu et al., 2014). On the national scale, the urban construction land expands more than six times in the last 30 years. For individual cities, urban proper also has been experiencing immense expansion. Beijing and Tianjin, as Chinese megacities, respectively expand from 801 km^2, 795 km^2 to 2452 km^2, 3343 km^2, increasing annually at 3.7%, 4.7% (Wu, 2015). The built-up area of Dongguan, a prefecture-level city in Guangdong Province, reaches 246 km^2 in 2003, which is eighteen times as large as it was in 1990 (Wu et al., 2006). Urban expansion of some small counties and towns is more extensive than the big ones: the growth rate of the built-up area of Chengguan Zhen in L county, Shandong Province was 3% per year from 1976 to 1999; but 20.6% annually between 1999 and 2003 (Han, 2010).

The rapid urban expansion characterized by low-density and excessive land consumption has caused severe problems. In the economic aspect, the continuing investment-driven growth results in increasing debt of the local authorities (He et al., 2006). In the social aspect, increasing cultivated agricultural land conversion generates fierce social conflicts and brings about social segregation (Zhou & Logan, 2008). The expansion of city proper prolongs the commuting distance and leads to mounting dependence on motor vehicles, triggering severe traffic congestion (Zhou & Logan, 2008). In the spatial aspect, housing vacancy in new town is juxtaposed with underutilized land use in the old town, manifesting the waste of land resources (Li et al., 2015). Ecologically, voracious energy consumption causes the degradation of urban environment and imperils food security (Chen et al., 2016). The growth-driven approach characterized by continuous urban spatial expansion has been proved unsustainable and impracticable.

3.3 Policy Reforms to Facilitate Urban Redevelopment at the Central Level

Two groups of central policies were proposed by the Chinese Central State to launch an urban redevelopment reform to address the severe issues caused by rapid urban development. One is to discourage low-density urban sprawl on non-urbanized land; the other one is to encourage intensified development with higher density in existing urbanized areas to improve the quality of urban development. Section 3.3.1 analyses the central policies to restrict the rural land transfer from the supply side. These policies play an essential role to motivate urban redevelopment in existing urbanized areas. The subsequent sections examined the central policies to address social resistance and facilitate urban redevelopment.

3.3.1 Institutional Control over Rural Land Conversion

To tackle the issues caused by rapid urban sprawl, the central state, from the supply side, establishes a series of policies to constrain the rural land conversion, as shown in Table 3–1. The year 2004 marked the first turning point of *Cun Liang Gui Hua.* By promulgating the *Decisions of the State Council on Deepening the Reform of Strict Land Management*, the State Council imposes institutional controls over the rural land transfer and encourage the revitalization of urban underutilized land in the built-up area (Lin, 2015). The policy arranged responsibilities between the central and local authorities—the central government should limit the quota supply for new urban construction land (*Zeng Liang Tu Di*) transferred from the greenfield while the local governments were authorized to revitalize the existing construction land of low intensity and inefficiency (*Cun Liang Tu Di*).

In the next year, to better instruct and motivate the local government, the central state listed the conservation of rural land into the basic national policy as "Saving Resources" (The State Council, 2005). For institutionalizing the constraint of agricultural land conversion, in 2008, the State Council established a strict rural land conservation scheme with relevant regulations and technical standards regarding the control of urban scale and intensification of existing construction land (The State Council, 2008).

In 2009, the Ministry of Land and Resources put forward a "most strictly ever" arable land red line to delineate the minimum scale of arable land (Ministry of Land and Resources, 2009). 120 million ha of greenfield are designated as permanent agricultural land to protect food production safety in China. These above policies institutionalize the protection of agricultural land and set stringent limitations on the rural land acquisition from the supply side. The importance of urban redevelopment has thus been underlined (Table 3.1).

Table 3.1 Rural land supply control policies at national level

Year	Policy statement	Policy implications
2004	Decisions of the state council on deepening the reform of strict land management (state council)	This policy highlighted the conservation of rural land in the urbanization process and encouraged the revitalization of urban underutilized land in the built-up area.
2005	Notices of the state council on doing a good job in constructing a conservation-oriented society (state council)	The central state brought the conservation of rural land into the basic national policy as "saving resources".
2008	Notices of the state council on promoting land intensive use (state council)	The policy required local governments to obey a strict rural land conservation institution, control the scale of urban land, make full use of existing construction land and establish policies as well as technical standards for the guidance of the intensification of land use.
2009	The arable land minimum policy (Ministry of Land and Resources)	This policy set limitations for the further acquisition of rural land in China and encouraged the intensive use of construction land from the supply side.

3.3.2 Property Right Reform in Urban Redevelopment

Since the country transformed from centrally planned economy to market economy, the pure public ownership was diversified into the coexistence of multi-ownerships. Simultaneously, the recognition and protection of property rights has become a pressing issue. From 2004, the central state commenced launching the property right reform to reduce the possibility of political influence on privatized property rights, especially in urban (re)development (Po, 2011). In 2004, the amendment to the 1982 Constitution conceded that when residents' properties were expropriated, they had the power to obtain fair compensation (He & Wu, 2009). In 2007, the central government released the Property Right Law to formally grant the "inviolability" of private property right to the property owners (National People's Congress, 2007). This law authorizes the property owners the power to possess, use, dispose or benefit from their properties. State's power of intervention has been restricted since the expropriation or requisition of private property has to aim for public interest (Zhang, 2008). In terms of the individual house expropriation, those who implement the expropriation shall provide fair compensation and relocated housing with the affected residents to ensure their living condition.

The issuance of property right law not only legalized the rights and obligations, but also conceptualized the cultural and moral notion that people have right to oppose the development they perceive as unfair (Abramson, 2011). The difficulties in assembling private ownerships in the built-up area hence have been escalated given the complexity of property rights and the increased diversity of stakeholders' interests (Po, 2011). However, the property right reform sets no specific requirement or procedures to guarantee a just implementation process for the housing expropriation, and the definition of the public interest is still ambiguous (Zhang, 2008). The lack of public participation and the consequently distributive inequality in the state-led redevelopment triggered increasingly fierce resistance from the residents who are empowered by the clearly defined property right (Mertha, 2009). The immense social protestation has been the most significant barrier to the proceeding of urban redevelopment (Lin, 2015). Since the establishment of policies to constrain the conversion of rural land, urban redevelopment has gradually become the primary approach for urban development, especially in megacities in which acute land scarcity in established areas hinders redevelopment (Ye, 2014). To facilitate the implementation of urban redevelopment, institutional reform to engage and share interests with the affected residents is necessary.

3.3.3 Participatory Redevelopment Reform to Permit Direct Land Transfer between Current and Potential Land Users

Facing the acute shortage of land resource in China, the Guangdong province plays a pioneering role in policy experiment to intensify its built-up area by urban redevelopment (Ye, 2011). To optimize land resources allocation and usage, improve urban living conditions and promote sustainable development in urban area, in 2009, under the agreement with the Ministry of Land and Resources, the Guangdong government established the *Sai Jiu Gai Zao* policy for the experiment of a market-led approach to facilitate urban redevelopment (Guangdong Provincial Government, 2009; Zhu, 2008).

This policy announced five fundamental principles to facilitate an "open, fair and equitable" urban redevelopment process. Firstly, this policy encourages the local government to play a conductive and regulative role in the process while promote the private developers to assume the leading operator of redevelopment projects. Secondly, it requires that all the land or properties that enlisted in redevelopment agendas should be legally registered and owned by the owners. In this case, the interests of the original property/landowners shall be protected with proper compensation. Thirdly, the land which is applicable to urban redevelopment should be consistent with the existing master plans. Fourthly, it attempts to resort to the market power to maximize the return of land resources in urban redevelopment. Fifthly, the urban redevelopment process should be fair and equitable and consistent with relevant laws and policies. According to the policy, three principal kinds of underutilized land are identified, including underutilized industrial sites, old town and urban villages.

To motivate original land users to engage in urban redevelopment, the policy restricts the dominance of the local government, delegates certain powers to existing land users and reallocates the land value increment between the state and existing landowners (Tian, 2018). It waives the traditional rule that land use transfer for commercial use has to go through a process in which local government assembles fragmented land use rights from the sitting users and auctions the acquired site to developers for redevelopment. The direct land use right transfer between the new land users (normally private developers) and the existing land users is allowed. The original property owners thus have the power to choose, negotiate and share interests with the involved developer in the transfer process (Lin, 2015).

Through the institutional reform of *San Jiu Gai Zao*, the central government works as an external force from the upper level to overcome the institutional separation and stimulate cross-boundary cooperation at local level (Schoon & Altrock, 2014). Municipal governments have had to redirect the development approach from promoting continuous spatial expansion to leading a fair, interest-sharing scheme of a spatial upgrade in specific area (Lin, 2015). The communities are empowered to join in the decision-making process and share the land conveyance income previously monopolized by the state (Li & Liu, 2018; Schoon, 2014b).

This reform is perceived as a starting point of maturing legal guidance for urban redevelopment and a core mechanism of China's reform since 1978 (Schoon, 2014a). It has important policy implications for the rest of China (Tian & Yao, 2018). In 2016, the Ministry of Land and Resources established a central policy to promote the market-oriented mechanism to other provinces in China. To motivate more stakeholders to join in the redevelopment process, it mainly emphasizes the role of market mechanism in interest redistribution and diminishes the leading role of the local government in the urban redevelopment process.

3.3.4 Changing Demolition and Relocation Policies to Ensure Fair Compensation

With an ambiguous definition of the public interest in the property right law, the growth-driven coalition between the local government and the developer facilitated the continuous enforcement of land acquisition and displacement through the demolition permit institution. This formal institutionalized procedure resulted in immense social conflict and unrest in urban redevelopment. To address, in 2011, the state council issued a new policy to abolish the demolition permit institution which restrains the intervention power of the local government (The State Council, 2011). Three principals have been released as just procedure, fair decision-making process and transparent outcome.

To ensure the procedure justice, the policy clearly conceptualizes public interest[2]and stipulates that compulsory housing expropriation only applies to public

[2] The public interest includes (1) the needs of national defence and diplomacy; (2) the need for infrastructure construction, such as energy, transportation and water conservancy, which is organized by the government; (3) the needs of public utilities such as science and technology, education,

interest purpose and only the local government should take charge of the acquisition process. Before the inception of the expropriation, the local government shall appropriately compensate the affected residents. In the implementation process, the actor authorized by the local government to conduct demolition cannot aim for profits and forceful strategies to facilitate the relocation are forbidden such as violent and threatening behaviours, energy supply cutting-off of water, heating, gas and electricity, road access obstruction and other illegal means.

In terms of the fair decision-making process and transparent outcome, the policy encourages more public participation in the formal institutional arrangement. It delegates the local government to make a relocation and compensation plan to ensure that the relocated residents receive financial compensation comparable to the market price of their properties and be entitled to temporary housing assistance (Ye, 2011). Once the plan is finished, the local government is required to hold a public hearing to publicly solicit comments from the affected residents. The revisions to the plan in accordance with the collected opinions are essential. Residents who oppose the revised proposal can appeal to the court for an administrative review. Once the final decision of expropriation is made, although the local government is empowered for the compulsory acquisition, it shall provide the appropriate compensation with the uncooperative residents. The new national regulation shapes a fair, transparent and just process and restructures the power deployment in the residential redevelopment (Ye, 2014).

3.3.5 *Policies to Facilitate Public-Private Cooperation in Residential Redevelopment*

To promote the residential redevelopment in built-up area, the Chinese government established a new approach to encourage the cooperation between the public and the private sectors. Since overcrowded and dilapidated residential areas resulted in underutilized land in the city centre, since 2013, through a series of planning policies, the central government has established a Public-Private Partnership mechanism to regenerate these shanty areas (*Peng Hu Qu*) in inner city.

In 2013, the State Council promulgated a policy to formally institutionalize the Shanty Area Regeneration (*Peng Hu Qu Gai Zao*) in China (The State Council, 2013). Under the principle that "government-led and market operation", this mechanism underlines the cooperation between public authority and market sector: the municipal government plays a guiding role in the implementing process, including

culture, health, sports, environment and resource protection, disaster prevention and mitigation, cultural relics protection, social welfare, municipal public administration and so on; (4) the need for the construction of affordable housing projects organized by the government; (5) the need for the alteration of the old city by the government in accordance with the relevant provisions of the urban and rural planning law, such as the concentration of dangerous buildings and the backwardness of infrastructure; (6) the needs of other public interests stipulated by laws and administrative regulations.

policy guidance, financial support and interest distribution mechanism establishment and the mobilization of affected residents and private developers. To further boost the implementation of the residential redevelopment, the State Council establishes a progressive policy in 2014 to require the local government to simplify the procedures for the project approval (The State Council, 2014). It also regulates that in the implementation process, the interest of the affected residents shall be appropriated protected.

In 2015, to motivate the involvement of the market sector, the State Council put forward an innovative approach—"government purchase" (The State Council, 2015). In this approach, the local government assumes a role to supervise and coordinate the urban redevelopment process and it is able to purchase the services from the private developer, in terms of housing expropriation and demolishment, relocation housing construction, monetary compensation and public infrastructure construction. These central policies establish a policy framework for the local policies, which significantly encourage the implementation of the Public-Private Partnership mode at local level.

3.4 Administrative Interventions to Facilitate Urban Redevelopment in Shenzhen

In line with the efforts from the central state and the provincial government, the Shenzhen municipal government promulgated its own policy framework to facilitate participatory urban redevelopment at local level (Fig. 3.1). Firstly, an array of policies was promulgated to constrain the conversion of rural land for containing urban sprawl. Secondly, the market-oriented mechanism was established to motivate market power to support the bottom-up redevelopment process, in which both market sectors and property owners are able to initiate the project for interest sharing and property right protection. Thirdly, the municipal government also regulated a state-led redevelopment approach to maintain the state power and supplement the market mechanism. Fourthly, facing pervasive poor performance of the developer-led residential redevelopment, the municipal government abolished the market-oriented mechanism and reversed to a government-led mechanism. Fifthly, the municipality established a Public-Private Partnership approach to further experiment the appropriate residential redevelopment mechanism. This section analyses the evolutionary process of participatory urban redevelopment reform in Shenzhen.

3.4.1 Policy Framework to Contain Urban Sprawl in Shenzhen

The built-up area of Shenzhen expanded very rapidly after it became a special economic zone (Fig. 3.2) (Xie, 2007). The total urbanized land was only 2.81 km^2 in

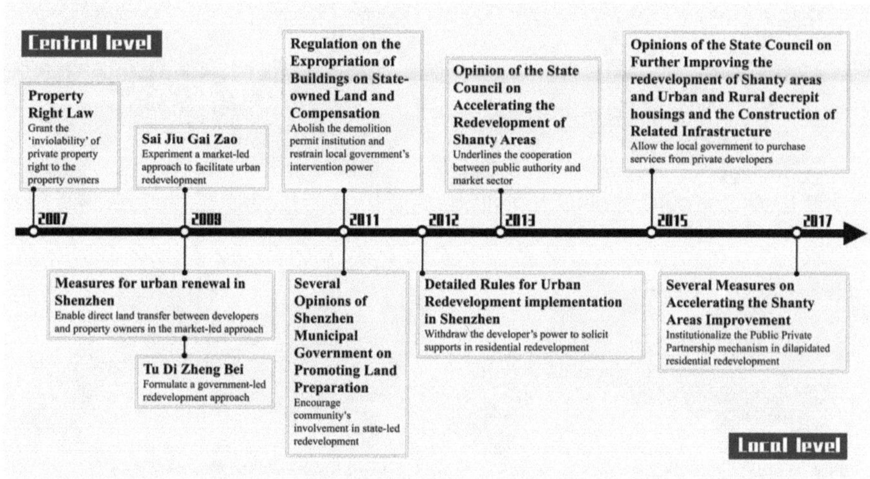

Fig. 3.1 Timeline of participatory redevelopment policies

1979, which increased to 139 km^2 by 1990. From 2000 to 2005, the average annual growth rate of the built-up area reached more than 45% (Xie, 2007). The total urbanized land area reached beyond 930 km^2 in 2005, covering almost 47% of the city size (Qian et al., 2015). The rapid development process resulted in excessive waste of land resources and pervasive underutilized land in the built-up area, which has severely harmed the sustainable development of social economy (Chen et al., 2016; Shi & Fan, 2011).

To improve the quality of urban development and keep consistent with the regulations at central level, the Shenzhen municipal government put forward its policy framework to constrain the rural-to-urban land conversion. To protect the ecological safety, in 2005, Shenzhen city government delineated a basic ecological control line as its urban growth boundary. With the control line, an ecological protection area[3] of 974 km^2 has been set in which any construction within the boundary is forbidden except necessary transport infrastructure and public facilities. This policy, from the supply side, strictly constrains the rural-to-urban land conversion and facilitates the protection of natural environment.

As a consequence of the strict greenfield conservation, the shortage of available rural land for urban growth has been intensified in Shenzhen. The municipal government commenced focussing on the revitalization of urban construction land in inner area and hindered the enhancement of city property expansion in periphery area. The year 2007 marked a turning point of urban development planning because of the establishment of the "draft master plan of Shenzhen (2007-2020)" (Shenzhen Planning and Land Resources Committee, 2007). In this master plan, the Shenzhen municipal government firstly highlighted the concept *Cun Liang Gui Hua* as the primary strategy

[3] The ecological protection area includes water, green spaces, nature reserves, farmland, forests and so on.

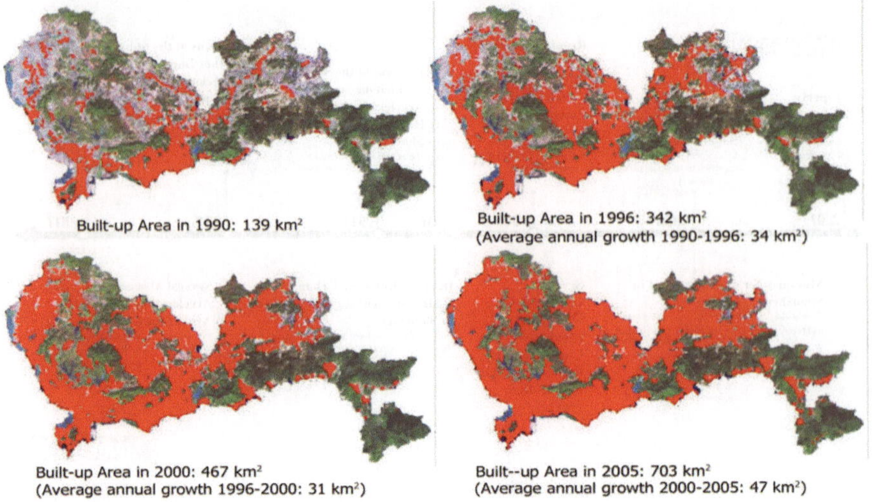

Fig. 3.2 Built-up area of Shenzhen in 1990, 1996, 2000 and 2005 *Source* Shenzhen Municipal Bureau of Urban Planning

to guide urban development (Shenzhen Planning and Land Resources Committee, 2007). *Cun Liang Gui Hua* refers to the efforts of using planning policies to encourage higher-density development in existing urbanized areas and discourage low-density urban sprawl in non-urbanized land (Han & O'Connor, 2008). With limited land resources for continuous urban expansion, Shenzhen regards urban redevelopment as a new opportunity to modernize the built environment, open up new spaces to enhance capital accumulation and sustain investment rather than maximize land revenue (Wu, 2018). Through urban redevelopment, high value-added land use would replace the underutilized land use; modernized community facilities, infrastructure and public housing would be constructed; and the urban environment would be eventually improved. Urban redevelopment thus becomes a significant approach to facilitating the economic and social transformation, the industrial structure optimization and land-use efficiency improvement in the built-up area in Shenzhen (Liu, 2010).

3.4.2 Urban Renewal Policy to Encourage Bottom-Up Urban Redevelopment from 2009 to 2012

To improve process efficiency and address social resistance in urban redevelopment, Shenzhen, a pioneer city in China's economic reform, has contributed its effort in policy innovation. In accordance with the *San Jiu Gai Zao* provincial policy, in 2009, the Shenzhen Municipal Government promulgated a landmark policy—"Measures for urban renewal in Shenzhen" to initiate the market-oriented reform of urban redevelopment (Shenzhen Municipal Government, 2009). This policy established a "market mechanism" to encourage the revitalization of its underutilized land. According to this mechanism, local government's role changed to a "night watchman" with minimized intervention, while land use transfer is directly conducted between developers and the affected residents through equal negotiation. The government merely focuses on policymaking and project approval while market sectors take charge of land assembly. On completion of the land assembly, the developer is required to pay a discount land-use fees to the local government, which is far lower than the market price and also has to provide social rental housing and community facilities to the local government (Schoon, 2014b). After property assembly and land-use fees payment, the developer takes charge of demolition and reconstruction to accomplish the redevelopment project.

Four kinds of underutilized land are identified in the redevelopment policy in Shenzhen: urban village, old industrial area, old town and dilapidated residential area. The policy clearly states only areas with valid and clear property rights are allowed to proceed to redevelopment agenda; residence conditions should also be verified as a shortage of necessary infrastructure and public facilities, harsh living environment or major security breach (Shenzhen Municipal Government, 2009).

This approach was intended to deploy greater market power in functioning of improved resource redistribution among the involved stakeholders. Before the endorsement of the participatory urban redevelopment policy, the land assembly process was monopolized by the local government in which the government assembled fragmented land use rights from the sitting users and auctioned the acquired site to developer for redevelopment (Yeh & Wu, 1996). Through this process, the local government and the developer formed a coalition to take control of the decision-making process and gain a large portion of the land value increment. The affected residents were excluded from the decision-making process, causing concern of distributive inequality. This mode of interest redistribution was increasingly concerned as unfair through open expression of local residents' dissatisfaction and incidents that sometimes threaten social stability. To address the distributive inequality, the participatory urban redevelopment policy regulates that with fixed proportion of the interests handing over to the local government, the rest of land value increment is designated to reallocate between the two contracting parties—developers and property owners. It motivates developer's investment and owners' participation to facilitate the implementation of urban redevelopment.

The policy also institutionalizes public participation by establishing a two-round participation mechanism. This mechanism was established based on a valid demolition permit scheme. In the first round, the developer should obtain support from more than two in three of the affected residents to become the candidate. On completion of that, the developer has to sign agreement with the majority of the affected residents (normally more than 90%) to apply for the demolition permit scheme. After retaining the permit, the developer is able to conduct compulsory acquisition from the last uncooperative residents to finalize the land assembly. However, the change of national policies has a significant impact on the policy implementation at local level. In 2011, the central government abolished the demolition permit scheme to pacify the social resistance caused by the conflicts between forced demolition right and private property right protection (Ye, 2014). The policy stipulates that compulsory land acquisition only applies to public interest purpose and only local government should take charge of the compulsory acquisition process (The State Council, 2011). When compulsory land acquisition occurs, all affected property owners and residents should be appropriately compensated.

The removal of the demolition permit scheme created major challenges for assembling property rights in urban redevelopment projects in Shenzhen, particularly in dilapidated residential redevelopment projects where property rights are fragmented and owned by individual household. In 2012, in line with the central policy, the Shenzhen municipal government established a new policy to amend the legitimate procedure of the residential redevelopment and withdraw the developer's power to solicit supports and make relocation and compensation scheme in the residential redevelopment (Shenzhen Municipal Government, 2012). However, in the approved eight DRAR projects, the developers had to obtain 100% consent from affected property owners to commence the demolition and relocation. The contradiction between the regulation and the implementation process extremely escalated the difficulty of the property assembly. The market-led residential redevelopment projects all suffered from severe delay or failure (Ma, 2013; Zhou, 2016). The market mechanism failed to facilitate successful bargaining between all existing and potential stakeholders which resulted in severe inefficient resource allocation and poor project performance.

3.4.3 Land Preparation Policy to Regulate State-Monopolized Land Acquisition in 2009

To improve the process efficiency of the construction of urban infrastructure, in 2009 Shenzhen formulates a government-led redevelopment approach—land preparation (*Tu Di Zheng Bei*) and piloted it in some experimental cases (e.g. Ludancun redevelopment). Apart from the urban renewal policy that aims to stimulate the market power, in the land preparation approach, the municipal government is authorized the power to make decision, arrange funds and negotiate with the affected property owners for land assembly. On completion of the property assembly, the municipal government was required to delegate a land clearance operator which cannot aim for profit in the implementation process. Once the land is cleared and prepared, the municipal government stores up the land into its land coordination and reserve centre for major infrastructure construction. As a supplement to the market mechanism, this approach is mainly based on the traditional housing acquisition mechanism to maintain the state power on urban land (Interview of WDB, Dec-28-2016). The state power to compulsorily withdraw the property or land from the original owners was maintained, through purchase or expropriation. Scholars noted that this government-led institution is designated to assemble and store up scattered underutilized land in which the developers are not interested, as market sectors always tend to exclusively engage in the redevelopment projects of premium location and superior profit (Xu et al., 2016). However, despite some pilot cases, the formal procedure of land preparation was not established in 2009.

Different from the traditional state-dominated land acquisition process, the institutional arrangement established mechanisms to encourage the involvement of the community in the decision-making process. In 2011, the Shenzhen municipal government formulated a policy to authorize the local government to take charge of the funding provision, plan-making, land expropriation (purchase) and building demolition in the land preparation process (Shenzhen Municipal Government, 2011). However, the abolishment of demolition permit scheme at central level in 2011 significantly constrained the power of the government in the housing expropriation process (The State Council, 2011). To make the local policy consistent with the upper policy, in 2013, Shenzhen formally institutionalizes the government-led housing expropriation scheme. The housing expropriation has to comply with the public interest while the affected residents have to be appropriately compensated through monetary compensation, property replacement or a combination of both approaches. Excluded are the real estate and property management companies which aim to seek profits in the expropriation process. This policy contributes to the establishment of a fair and transparent procedure for the state-dominated urban redevelopment approach.

3.4.4 Cessation of the Market-Oriented Residential Redevelopment Approval since 2012

In line with the central policy, in 2012, the municipality replaced the market-led residential redevelopment approach with a government-led residential redevelopment mechanism (Shenzhen Municipal Government, 2012). The authority of agenda setting transfers from the market sector to the district government regarding supports solicitation, redevelopment unit plan establishment and relocation and compensation schemes making. On the completion of the approval, the district government takes charge in establishing relocation and compensation schemes and negotiating with the affected residents. After more than 90% of residents signed agreements, the district government publicly auctions the redevelopment project to a private developer for further property assembly and implementation. When the developer obtains 100% consents from affected property owners, it is able to demolish and reconstruct the residential area.

At the end of 2012, the municipal government held a secret meeting with all the relevant departments to halt residential redevelopment. According to interview, "*we were invited to a private meeting in the municipal government. In the meeting, the municipal government claimed that the large residential areas which were designed and constructed in the 1980s and 1990s would not be considered for redevelopment in five years apart from major safety hazards of the buildings or the requirement of major infrastructure and public facilities. As the approved market-led residential redevelopment projects all were suffering from serious project delay, the municipal government suspended the residential redevelopment approval through the adoption of the government-led mechanism in 2012.*" (Interview of YKL, Jan-12-2017).

After the meeting, the municipal government promulgated an internal document to the involved departments to further clarify the government's attitude towards residential redevelopment (Interview of an official in Nanyuan Street office, Jan-12-2017). According to the document, the municipal government suspended any residential redevelopment applications and prohibited the private developers' engagement in the dilapidated residential areas. It also set an upper limit of the Demolition Compensation Ratio (DCR) as 1:1.3 for the approved residential redevelopment projects to curb opportunistic behaviours of the affected residents. For residential areas with inferior residence conditions, the government commenced encouraging a reinforcing approach to protect the residents' safety. However, merely through building structure reinforcement and wall cracks repair, the reinforcing method failed to improve the poor building quality and living conditions substantially, so it triggered fierce opposition from the affected residents.

3.4.5 Shanty Area Improvement Policy to Encourage Public-Private Partnership Redevelopment since 2016

In accordance with the central policy, since 2016, a new mechanism of residential redevelopment—shanty area improvement (*Peng Hu Qu Gai Zao*)—has been introduced in Shenzhen. Different from the dichotomization of market-led and government-led, this approach encourages the cooperation between the local government and market sector to tackle the difficulties of property assembly in the residential redevelopment process. The local government takes charge of the support solicitation and then purchases services from the private developers in terms of compensation and relocation schemes establishment, agreement signing, demolition and construction.

In January 2017, to facilitate the implementation of urban redevelopment, the municipal government delegated the authority of redevelopment approval and supervision to the district government in Shenzhen (Gan, 2017). In August 2017, the Shenzhen municipal government promulgated a draft regulation to institutionalize the Public-Private Partnership mechanism in the dilapidated residential area redevelopment (Shenzhen Housing and Construction Bureau, 2017). In this mechanism, the responsibility of property right confirmation, living condition examination and redevelopment support solicitation transferred from the district government to the street office, the grassroot governmental institute. On completion of the above tasks, a project temporary organization will be set on site on behalf of the district government. Developers compete to win the operator through a public bidding process and then cooperates with the street office and the affected residents to formulate the compensation and relocation schemes. For the schemes, the ceiling of the DCR has been established as 1:1.2 to constrain residents' extremely high expectations. To ensure the shanty area improvement aiming for public interest, the newly constructed apartments are all dedicated to in-situ relocation housing and public housing. In the property assembly, with the agreements from the majority (more than 95%), the district government is able to compulsorily purchase or expropriate the houses from the remaining uncooperative residents.

As the governmental official indicated, "*the shanty area improvement is a combined redevelopment model between the government-led and the developer-led approaches. The removal of the 100% agreement regulation is supposed to greatly facilitate the property assembly process in the residential redevelopment.*" (Interview of YKL, Jan-12-2017) In June 2018, the draft policy of the Public-Private Partnership residential redevelopment mechanism was formally approved (Shenzhen Municipal Government, 2018).

3.5 Changing Approaches to Dilapidated Residential Redevelopment in Shenzhen

In line with the participatory urban redevelopment policies at central level, the Shenzhen municipal government promulgated an array of policies to institutionalize the residential redevelopment into three simultaneous mechanisms: government-led redevelopment mechanism, developer-led redevelopment mechanism and owner-led redevelopment mechanism. These mechanisms endeavour to address the difficulties in property assembly in the residential redevelopment by restricting the power of the resourceful stakeholders, such as the local government and the private developer and empowering and sharing interests with the traditionally powerless stakeholder—the affected residents. This section illustrates the three residential redevelopment mechanisms.

3.5.1 Government-Led Approach Established by Land Preparation Policy

In line with the central policy, in 2013, the Shenzhen municipal government establishes the formal procedure of housing expropriation as the formal institutional arrangement of the government-led residential redevelopment approach (Fig. 3.3). To constrain the state power and protect the interest of the affected residents, the policy establishes several requirements to regulate the government's behaviours.

At first, the housing expropriation department shall determine the specific scope of the project according to the property rights and residence conditions. After that, the related department shall publish the expropriation notices to the community as well as the public both online and on site. On completion of scope determination,

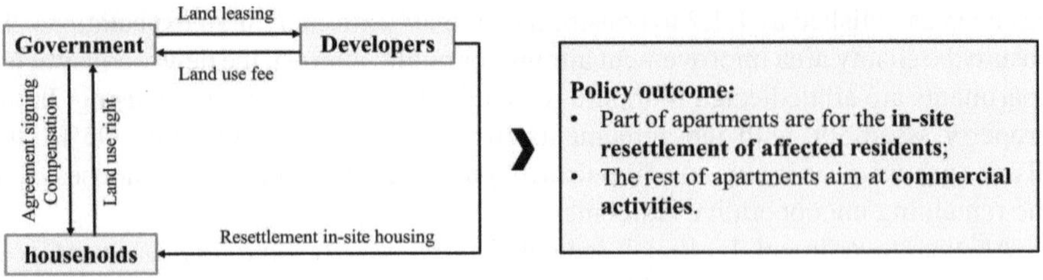

Fig. 3.3 Institutional arrangement of government-led redevelopment

the department shall investigate on the ownership, use, area of the affected houses and formulate a budget plan for the redevelopment. Based on the initial investigation and budget plan, the housing expropriation department shall establish a compensation scheme and submit it to the district government. The scheme should include the approaches and standards of compensation, the location and area of the relocation housing, rewards to households who move out in advance, and subsidy for relocation and rental houses, etc. The district government is required to discuss the compensation scheme with the financial, development and audit departments. After that, the district government is obliged to publish the scheme for public consultation. The affected residents who contradict to the scheme can submit feedback to the housing expropriation department. The local government is required to revise the compensation scheme according to the collected feedback and publish online without delay. If more than half of the affected property owners or the property owners who hold more than half of the areas of the community consider the compensation scheme is contradictory to the house expropriation regulation, the municipal government shall continue to revise the scheme. The compensation scheme shall be passed if it obtains the consent from the majority. Otherwise, the municipal government must suspend the housing acquisition. On completion of the compensation scheme, the municipal government promulgates a determination of house expropriation,[4] which signifies the end of consultation and the start of the implementation process.

Residents shall be appropriately compensated in the following perspectives: in-kind housing or monetary payment compatible with the value of the expropriated house[5]; subsidy for the relocation and contemporary rental house; allowance for the losses of production or business suspension caused by the acquisition of houses before the building demolition. Upon the implementation of building demolition, the housing expropriation department shall conclude an expropriation compensation agreement[6] with the affected property owners. Forbidden are violence, threats, the

[4] The determination includes the name of the project, spatial scope, compensation scheme, expropriation implementation organization, period of house expropriation, reception location and supervision approach

[5] To protect the fair evaluation of the house value, the compensation price of the expropriated house shall be assessed and determined by a qualified real estate appraisal agency.

[6] The content of the expropriation compensation agreement shall include: compensation method, compensation amount and time limit for payment, place and area for the exchanged houses,

interruption of the water, heating, gas and power supply, road access or other illegal methods to force the affected residents to relocate. The government-led mechanism contributes to the protection of interest protection of the affected residents.

3.5.2 Developer-Led Approach Framed by Urban Renewal Policy

The market-led residential redevelopment mechanism establishes a two-round public participation scheme in the property assembly process to actively engage the community affected by the redevelopment process (Fig. 3.4). After an investigation of the residents' demands and an evaluation of project investment and profit, the developer begins to promise newly constructed apartments and compensation to obtain supporting documents (printed copies of ownership certificates and authorization letters) from the affected residents. On the completion of more than two-thirds of the documented supports, the developer submits a project application to the local government and becomes a candidate of redevelopment upon project approval. The local government and developers exchange information to verify the property ownership and unit quality. The regulation clearly states that only residential areas with valid and clear property rights are allowed to submit a redevelopment agenda; residence conditions should also be verified by marking out shortage of necessary infrastructure and public facilities, harsh living environment or major security breach (Shenzhen Municipal Government, 2009). Once the project is approved, the developer commences to apply for an approval of their redevelopment plan including the supply of land for public infrastructures such as roads, parks, open spaces and parking areas and potential floor area ratio (FAR). The local government assesses and approves their redevelopment proposal including the FAR, land contribution and the requirement of public facilities (Shenzhen Municipal Government, 2012).

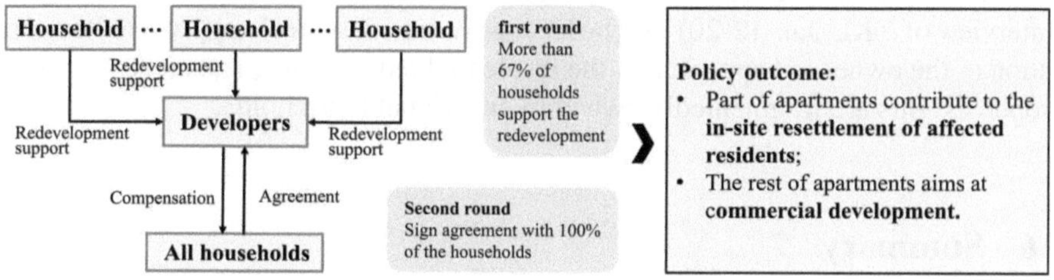

Fig. 3.4 Institutional arrangement of developer-led redevelopment

delivery time, relocation fee, temporary resettlement fee, or turnover settlement, loss of production suspension, property right cancellation method, relocation period, transitional methods, transition deadlines, related incentives or subsidies.

The second round is for the developer to organize consents with relocating residents. The developer takes values and goals of the affected residents into consideration and establishes a detailed compensation agreement including the decided compensation approach, a fixed DCR, amount of the compensation, the payment term, area and location of the relocation apartment and other relevant details. According to the regulation, the developer will be able to obtain demolition permit from the local government for compulsory acquisition if they can achieve at least 90% approval rate from the residents. This two-round public participation created new structural power to the residents (Ye, 2011, 2014). After the abolition of demolition permit scheme at the central level, the developer had to obtain 100% agreements from the affected property owners to commence the demolition and relocation in the approved market-led residential redevelopment.

3.5.3 Owner-Led Approach Formed by Urban Renewal Policy

To protect the interests of the property owners, the Shenzhen municipal government also encouraged self-motivated redevelopment in the market-oriented mechanism. This "owner-led" redevelopment also complies with the two rounds of public participation procedure as in the developer-led mechanism (Fig. 3.5). But the initiator in the first-round changes to be the affected residents. In the first round, property owners can take charge in soliciting redevelopment support. If more than two-thirds of property owners in one specific site agree with the demolition and construction, the whole property owners can be considered a group. This group is authorized to invite and cooperate with one private developer to apply the redevelopment agenda with the local government. In the second round, on the completion of the approval, the developer is required to establish the compensation and relocation schemes and obtain consents from the affected property owners. This approach is designated to facilitate the redevelopment of urban villages, in which the land is in charge by a powerful village collective while buildings are owned by each individual household (Interview of YKL, Jan-12-2017). The local government scarcely approved the application of the owner-led approach in the residential redevelopment projects, in which property rights are fragmentedly owned by individual households.

3.6 Summary

This chapter introduces the institutional context of urban redevelopment in order to clarify how central and local policies constitute key exogenous influences and how specific residential redevelopment mechanisms are generated at the local level. The market-oriented reforms of the 1980s triggered an accelerated urbanization process, accompanied by large-scale rural land conversion and rapid urban sprawl. This aggressive urbanization trajectory gave rise to severe urban problems and placed

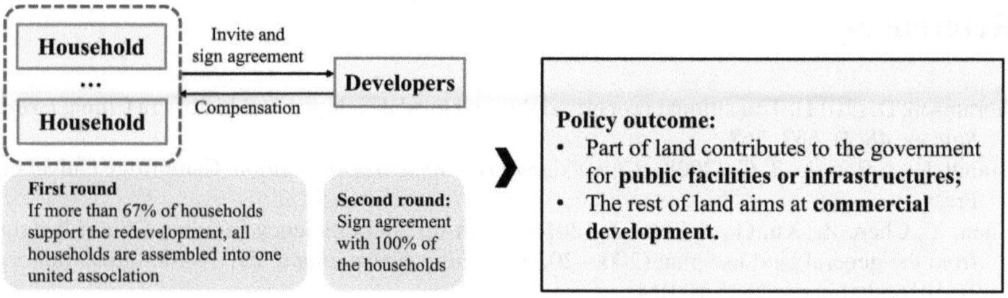

Fig. 3.5 Institutional Arrangement of Owner-Led Redevelopment

sustainable development at risk. In response, the central government introduced a series of policies aimed at curbing low-density urban sprawl in peripheral areas and promoting higher-density development within existing built-up areas. As a result, urban redevelopment has been strengthened as a core instrument for intensifying land use in pursuit of more sustainable urban development. To mitigate social conflicts and facilitate the implementation of urban redevelopment, central-level policies established key institutional mechanisms that both constrained the discretionary power of local governments and empowered residents to participate in decision-making processes. In Shenzhen, the municipal government translated these central policy directives into locally specific urban redevelopment regulations and practices. Through the interaction between central and local policies, three distinct residential redevelopment approaches have emerged: government-led, developer-led and owner-led (Fig. 3.6). The elucidation of these newly established residential redevelopment mechanisms provides the institutional foundation for the empirical analysis of the three redevelopment cases examined in the following chapters.

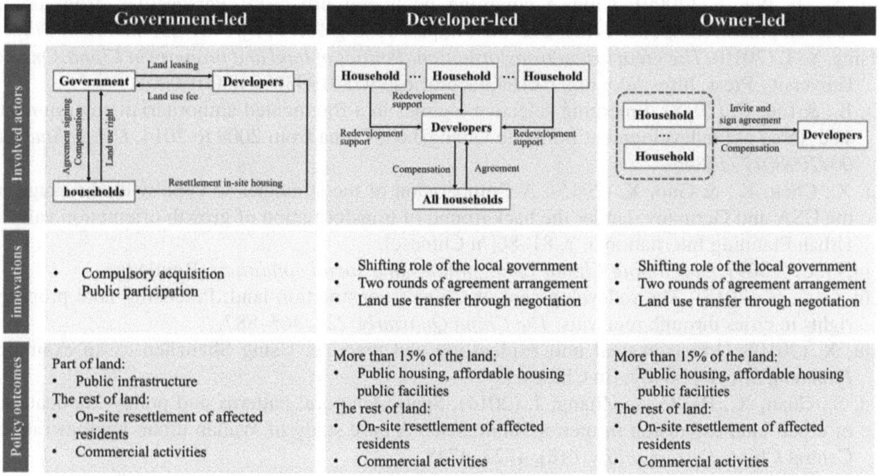

Fig. 3.6 Comparison of three institutional arrangements

References

Abramson, D. (2011). Transitional property rights and local developmental history in China. *Urban Studies, 48*(3), 553–568.

Brandt, L., & Rawski, T. G. (2008). *China's great economic transformation.* Cambridge University Press.

Chen, Y., Chen, Z., Xu, G., & Tian, Z. (2016). Built-up land efficiency in urban China: Insights from the general land use plan (2006–2020). *Habitat International, 51*, 31–38. https://doi.org/10.1016/j.habitatint.2015.10.014

Ding, C., & Knaap, G. (2005). Urban land policy reform in China's transitional economy. *Emerging land and housing markets in China, 2*, 32.

Dowall, D. E. (1993). Establishing urban land markets in the People's Republic of China. *Journal of the American Planning Association, 59*(2), 182–192.

Dowall, D. E. (1994). Urban residential redevelopment in the People's Republic of China. *Urban Studies, 31*(9), 1497–1516.

Fang, K., & Zhang, Y. (2003). Plan and market mismatch: Urban redevelopment in Beijing during a period of transition. *Asia Pacific Viewpoint, 44*(2), 149–162.

Gan, L. (2017). Authority decentralization from the municipal government to district government stimulates new vitality of the City. *Shenzhen Daily.*. http://sztqb.sznews.com/html/2017-01/10/content_3706031.htm

Guangdong Provincial Government. (2009). *Several Opinions of the Guangdong Province Government to Promote Three Old Redevelopment for Urban Land Use Consolidation.*

Han, S. S. (2010). Urban expansion in contemporary China: What can we learn from a small town? *Land Use Policy, 27*(3), 780–787. https://doi.org/10.1016/j.landusepol.2009.10.010

Han, S. S., & O'Connor, K. (2008). Urban consolidation and house prices: A case study of Melbourne 1990–2004. *GeoJournal, 73*(4), 285–295.

Han, S. S., & Wang, Y. (2003). The institutional structure of a property market in inland China: Chongqing. *Urban Studies, 40*(1), 91–112.

Harvey, D. (2002). From managerialism to entrepreneurialism: The transformation of urban governance in late capitalism. *The blackwell city reader*, 456–463.

He, S., Li, Z., & Wu, F. (2006). Transformation of the Chinese City, 1995-2005: Geographical perspectives and geographers' contributions. *China information, 20*(3), 429–456.

He, S., & Wu, F. (2009). China's emerging neoliberal urbanism: Perspectives from urban redevelopment. *Antipode, 41*(2), 282–304. https://doi.org/10.1111/j.1467-8330.2009.00673.x

Hsing, Y.-T. (2010). *The great urban transformation: Politics of land and property in China.* Oxford University Press. https://doi.org/10.1093/acprof:oso/9780199568048.001.0001

Li, B., & Liu, C. (2018). Emerging selective regimes in a fragmented authoritarian environment: The 'three old redevelopment'policy in Guangzhou, China from 2009 to 2014. *Urban Studies, 0042098017716846.*

Li, X., Chen, K., & Guo, X. (2015). A Comparision of the strategies to cope with shrinkage in the USA and Germany: Under the background of transformation of growth orientation values. Urban Planning International, 2, 81–86[in Chinese].

Lin, G. C. (2009). *Developing China: Land, politics and social conditions.* Routledge.

Lin, G. C. S. (2015). The redevelopment of China's construction land: Practising land property rights in cities through renewals. *The China Quarterly, 224*, 865–887.

Liu, X. (2010). Urban renewal unit exploration and practice: Using Shenzhen as an example. *Planners, 26*(11), 66–69. [in Chinese].

Lu, S., Guan, X., He, C., & Zhang, J. (2014). Spatio-temporal patterns and policy implications of urban land expansion in metropolitan areas: A case study of Wuhan urban agglomeration, Central China. *Sustainability, 6*(8), 4723–4748.

Ma, J. (2013, Jan–30). Who signed the supporting documents using my name? Developer was accused of forging supporting documents in Nanyuan Xincun. *Fang.com.* http://sz.esf.fang.com/newsecond/news/9462957.htm

Mertha, A. (2009). "Fragmented authoritarianism 2.0": Political pluralization in the Chinese policy process. *The China Quarterly, 200*, 995–1012.
Ministry of Land and Resources. (2009). *The arable land minimum policy.*
National People's Congress. (2007). *Property Right Law.*
Oi, J. C. (1992). Fiscal reform and the economic foundations of local state corporatism in China. *World Politics, 45*(1), 99–126.
Po, L. (2011). Property rights reforms and changing grassroots governance in China's urban—Rural peripheries: The case of Changping District in Beijing. *Urban Studies, 48*(3), 509–528.
Qian, J., Peng, Y., Luo, C., Wu, C., & Du, Q. (2015). Urban land expansion and sustainable land use policy in Shenzhen: A case study of China's rapid urbanization. *Sustainability, 8*(1), 16. https://doi.org/10.3390/su8010016
Schoon, S. (2014a). Chinese strategies of experimental governance. The underlying forces influencing urban restructuring in the Pearl River Delta. *Cities, 41*, 194–199. https://doi.org/10.1016/j.cities.2014.01.008
Schoon, S. (2014b). Three olds: Experimental urban restructuring with Chinese characteristics, Guangzhou and Shenzhen in comparison. In *Maturing megacities* (pp. 105–121). Springer.
Schoon, S., & Altrock, U. (2014). Conceded informality. Scopes of informal urban restructuring in the Pearl River Delta. *Habitat International, 43*, 214–220.
Shenzhen Housing and Construction Bureau. (2017). *Several Measures on Accelerating the Shanty Areas Improvement (Draft).*
Shenzhen Municipal Government. (2009). *Measures for urban renewal in Shenzhen.*
Shenzhen Municipal Government. (2012). *Detailed Rules for Urban Redevelopment implementation in Shenzhen.*
Shenzhen Municipal Government. (2018). *Implementation Opinions on Strengthening the Shanty Area Improvement in Shenzhen.*
Shenzhen Planning and Land Resources Committee. (2007). *Draft Master Plan of Shenzhen (2007–2020).*
Shi, A., & Fan, Z. (2011). Regulation on construction land scale from increment-led to revitalization-oriented. City Planning Review, 35(8), 88–96 [in Chinese].
Shin, H. B. (2009). Residential redevelopment and the entrepreneurial local state: The implications of Beijing's shifting emphasis on urban redevelopment policies. *Urban Studies, 46*(13), 2815–2839.
The State Council. (2005). *Notices of the State Council on Doing a Good Job in Constructing a Conservation-Oriented Society.*
The State Council. (2008). *Notices of the State Council on Promoting Land Intensive Use.*
The State Council. (2011). *Regulation on the Expropriation of Buildings on State-owned Land and Compensation.*
The State Council. (2013). *Opinion of the State Council on Accelerating the Redevelopment of Shanty Areas.*
The State Council. (2014). *Notice of the General Office of the State Council on Further Strengthening the Redevelopment of Shanty Areas.*
Tian, L., & Yao, Z. (2018). From state-dominant to bottom-up redevelopment: Can institutional change facilitate urban and rural redevelopment in China. *Cities, 76*, 72–83.
Wang, Y. P. (2001). Urban housing reform and finance in China: A case study of Beijing. *Urban Affairs Review, 36*(5), 620–645.
Wang, Y. P., & Murie, A. (1996). The process of commercialisation of urban housing in China. *Urban Studies, 33*(6), 971–989.
Wang, Y. P., Shao, L., Murie, A., & Cheng, J. (2012). The maturation of the neo-liberal housing market in urban China. *Housing Studies, 27*(3), 343–359.
Wong, C. P. (2000). Central-local relations revisited the 1994 tax-sharing reform and public expenditure management in China. *China Perspectives*, 52–63.
Wu, F. (2004). Residential relocation under market-oriented redevelopment: The process and outcomes in urban China. *Geoforum, 35*(4), 453–470.

Wu, F. (2018). Planning centrality, market instruments: Governing Chinese urban transformation under state entrepreneurialism. *Urban Studies, 0042098017721828.*

Wu, F., Xu, J., & Yeh, A. G.-O. (2006). *Urban development in post-reform China: State, market, and space.* Routledge.

Wu, J. (2015, December, 4). The price of Ludancun apartment is 15 thousand per square metre. *Real Estate Opinion Network.* http://www.guandian.cn/article/20151204/168484.html

Xie, L. (2007). A transitional city: the case study of Shenzhen, China, 1980–2005. In.

Xu, Y., Xue, Z., Wang, M., & Li, L. (2016). *The new land reform in Shenzhen.* China Citic Press.

Yang, Y.-R., & Chang, C.-H. (2007). An urban regeneration regime in China: A case study of urban redevelopment in Shanghai's Taipingqiao area. *Urban Studies, 44*(9), 1809–1826. https://doi.org/10.1080/00420980701507787

Ye, L. (2011). Urban regeneration in China: Policy, development, and issues. *Local Economy, 26*(5), 337–347. https://doi.org/10.1177/0269094211409117

Ye, L. (2014). Examining China's urban redevelopment: Land types, targeted policies, and public participation. In *Maturing megacities* (pp. 123–138). Springer.

Yeh, A. G., & Wu, F. L. (1996). The new land development process and urban development in Chinese cities. *International Journal of Urban and Regional Research, 20*(2), 330.

Yue, W., Liu, Y., & Fan, P. (2013). Measuring urban sprawl and its drivers in large Chinese cities: The case of Hangzhou. *Land Use Policy, 31*, 358–370. https://doi.org/10.1016/j.landusepol.2012.07.018

Zhang, M. (2008). From public to private: The newly enacted Chinese property law and the protection of property rights in China. *Berkeley Bus. LJ, 5*, 317.

Zhang, T. (2002). Urban development and a socialist pro-growth coalition in Shanghai. *Urban Affairs Review, 37*(4), 475–499. https://doi.org/10.1177/10780870222185432

Zhang, T. W. (2000). Land market forces and government's role in sprawl-the case of China. *Cities, 17*(2), 123–135. https://doi.org/10.1016/S0264-2751(00)00007-X

Zhou, S. (2016, Jan–14). Eight residential redevelopment cases in Shenzhen facing enornous difficulties in redevelopment implementation. *Shenzhen Evening Newspaper.* http://sz.southcn.com/content/2016-01/14/content_140771558.htm?COLLCC=1280225329&COLLCC=1949902161&COLLCC=1280225329&

Zhou, Y., Lin, G. C., & Zhang, J. (2018). Urban China through the lens of neoliberalism: Is a conceptual twist enough? *Urban Studies, 0042098018775367.*

Zhou, Y., & Logan, J. R. (2008). Growth on the edge: The new Chinese metropolis. Urban China in transition, 140–160 [in Chinese].

Zhu, G. (2008, 22 December). The Ministry of Land and Resources and the Guangdong Provincial Government signed a cooperative agreement to jointly build a pilot policy for economical and intensive use of land in Guangdong. *The Chinese Newspaper of Land and Resources.* http://www.mlr.gov.cn/zwgk/ldzc/xss/ldhd/200812/t20081222_113226.htm

Zhu, J. (2004). From land use right to land development right: Institutional change in China's urban development. *Urban Studies, 41*(7), 1249–1267. https://doi.org/10.1080/0042098042000214770

Chapter 4
Government-Led Participatory Redevelopment in Ludancun

4.1 Introduction

This chapter examines the power relations among the principal stakeholders—the local government, affected residents and developers—in a government-led redevelopment project. Section 4.2 introduces the case study project. Section 4.3 analyses the social practices and evolving power relations throughout the redevelopment process. Section 4.4 discusses the impacts of exogenous influences, while Section 4.5 identifies the effects of endogenous influences. Section 4.6 presents the spatial and social outcomes of the redevelopment and evaluates the associated power relations. Section 4.7 concludes the chapter.

4.2 Case Context and Community Attribute

Ludancun is a large and central-located dilapidated residential area with one kilometre away from the CBD of Luohu District, as shown in Fig. 4.1. It is situated by the Shenzhen River with the Mai Po Nature Reserve of Hong Kong on the other side. The area of Ludancun is about 66,473 square metres, with 1280 apartment units in 24 seven-storey buildings. The original FAR of the residential area was about 1.57. The premium location and great landscape view of the community escalate the commercial value of the land.

The old community was built by the municipal government in 1987 and completed in 1989. Upon its completion, Ludancun was the most upscale residential area in Shenzhen, which was awarded "the first civilized community in China" (Zhou et al., 2014). The housing was allocated as welfare housing to the well-educated workers who came to support the development of the special economic zone from all over the country. These staffs were working in the local government, public institutions and state-owned enterprises. Among the 1280 households, 756 were cadres in high

X. Li, *Participatory Redevelopment as Contested Arenas in China*,
https://doi.org/10.1007/978-981-95-9394-1_4

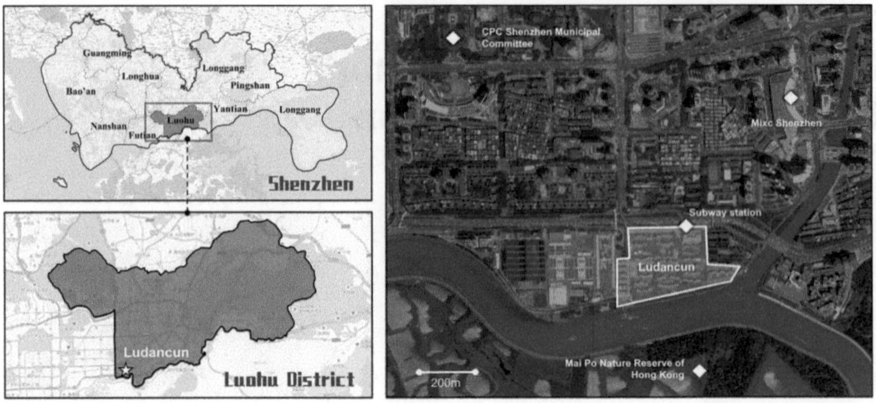

Fig. 4.1 The location of Ludancun, Shenzhen

positions in the local government, including 63 department-level, 227 division-level and 466 section-level cadres. Keren Chen, the head of the property owners committee and one of the earliest residents to move in Ludancun noted,

> It was very strenuous and competitive to get admission to live in Ludancun. Only the outstanding government cadres, scholars and engineers were able to live here. After several rounds of assessment, I got a superior score and was luckily awarded a three-bedroom unit. (SZhomeBBS, 2000)

However, with increasing residents moved into Ludancun, they were frightened by the poor construction quality. Mr. Lin who moved to Ludancun in 1990, one year after its completion, said,

> Before we moved in, the water pipes had been rusty, and some of them were even leaking. When you opened the tap, the water coming out was yellow. (Zhou et al., 2014)

After two years, the building quality became terrifying. Most of the external walls were cracking with water leaking in the rainy days. Inside the apartment, the roof was peeling off with severely corroded steel bars exposed. The leaking water resulted in a very damp environment inside the house, which even nurtured black mold.

According to Mr. Lin, *"...empty was inside the wall. When you were knocking it, it sounded like drums. Even nails cannot be stable in the wall."* (Zhou et al., 2014).

Why did the new buildings perform so inferior? According to Weiqing, who was a senior engineer participating the construction of Ludancun, the construction team used a large amount of sea sand in the structure. The sea sand cracked the wall and corroded the steels and pipes. In addition, the constructor used inferior and shoddy materials when building this residential area to reap an extreme profit. Cement, steel, brick and other raw materials used in Ludancun were far below the technical requirements.

With extraordinary dissatisfaction to the residential environment, residents had begun to complain to the local government since 1994. In 1996, the Shenzhen Housing Bureau dispatched professionals to investigate the building conditions in

Ludancun. After the inspection, the municipal government announced the building quality was stable and just utilized some waterproof material such as asphalt to fill the cracks. As the large mesh lines made the buildings look like "Zebra", residents began to call the housing "Zebra House". Besides the construction issues, the necessary public facilities were deficient. For example, there was no parking lot in the area and residents had to park on the road.

The residents living in Ludancun have strong social bonds, and most of them have a close social relationship with each other (Interview of HY, Jan-09-2017). When Ludancun was built, the property owner committee[1] (POC) was established. Some prestige residents were promoted as committee members. These residents were retired cadres in high positions. They voluntarily took charge of community management and maintenance. When residents attempted to negotiate with the government for Ludancun redevelopment, the POC assumed a substantial role to express residents' requirement to the municipal government. In 2007, a senior resident association[2] (SRA) was formed to organize public activities for older people in Ludancun. The SRA also played an important role in the resident mobilization in the redevelopment process.

4.3 Power Relations Analysis in Ludancun Redevelopment Project

The Ludancun redevelopment adopted a government-led approach and unfolded over a protracted period of more than 18 years, from 2000 to 2018. During this process, the local government adopted three governing strategies to secure the support and agreement from the affected residents. Accordingly, the redevelopment evolved through three temporally sequenced stages, each corresponding to a specific "relational power arenas": authoritative management, equal consultation and mandatory enforcement. These arenas were identified to exhibit dynamic power deployment among the main stakeholders, namely the local government, the affected residents and the developer.

4.3.1 Authoritative Management Arena

Out of the dissatisfaction with the poor living condition, residents sent petitions to the local government in the late 1990s. After receiving the petition of the affected

[1] Property owner committee (Ye Zhu Wei Yuan Hui) is a formal organization of property owners (Ye Zhu) for engaging a property service provider who organizes owners meeting and manages building maintenance in the residential area.

[2] Senior resident association (Lao Nian Xie Hui) is a voluntary group organized by the senior residents in Ludancun. They have regular activities to promote social bond and maintain mental and physical health of the senior people.

residents, the local government began to participate in Ludancun redevelopment. In 2000, based on its dominant power, the local government tried to negotiate with the residents for a mutually agreeable redevelopment approach and thus established the authoritative management arena. The major contradiction between these two parties was the distribution of responsibilities and benefits.

In June 2000, out of the annoyance at the poor living environment of Ludancun, more than sixty property owners jointly wrote an official complaint letter to the Municipal Government. Since the owners were high-level cadres in the government, the mayor responded quickly and paid a visit to the residential area in person to inspect the actual condition. During his visit, he witnessed the severe rainwater leaking into two families' homes, which urged him to take immediate action to tackle the deteriorating living condition (Zhou et al., 2014). The major made an official public announcement and said,

> This is not a superficial 'skin' issue but a terrible 'body' issue. Only 'comprehensive surgery' can solve the problem. Major redevelopment aiming at improving the living environment in this residential area will be taken seriously, with immediate effect. (Zhou et al., 2014)

The redevelopment plan that the mayor proposed was to demolish all existing buildings and build a new residential area in Ludancun. The decision, however, triggered intense internal debates within relevant government departments. At that time, the issue of deteriorating living condition had emerged in most residential areas built by the municipal government in the same period. As a residential area with the worst situation but most government attention, Ludancun redevelopment would become the first residential redevelopment project in Shenzhen. The Ludancun redevelopment approach would play an exemplary role for the following projects. *"The municipal government had to be very cautious about it"*. (Interview of WDB, Dec-28-2016).

The officials were at odds on how to fix the poor-quality issues and improve safety in Ludancun. Some supported the demolition and construction approach while some proposed to repair the existing buildings because the residential area just had been built for ten years. Despite the violent internal controversy, three months later, the municipal government decided to redevelop Ludancun. Interviews evidence demonstrated that the decision was made by the mayor (as an application of choice rule).

> With the instructions from the superior (the mayor), we, as subordinates, had no choice but to be cooperative…The reconstruction could not only improve the terrible living environment but also ameliorate the physical image of Shenzhen. These were the reasons for the mayor to make that decision. (Interview of WDB, Dec-28-2016)

Once the decision was made, the municipal government delegated the Municipal Housing Bureau to negotiate with the residents for the redevelopment plan and compensation schemes. To understand affected residents' demands, on November 23, 2000, the Housing Bureau sorted out more than one hundred officials to solicit opinions house by house on site (Feng, 2000). Upon the feedback collection, the Housing Bureau formulated a redevelopment plan and compensation schemes and directly submitted to the municipal government. According to the scheme, residents would be compensated for a newly constructed apartment unit with the same size and the subsidies for relocation and temporary rental houses (SZhomeBBS, 2000). As the government had limited financial resources, the original residents were required to share the cost of the project: each household had to disburse around 100,000 RMB. As an official in the Housing Bureau said,

> The only accessible funding was the housing funding, which was collected from individual households in Shenzhen for their community maintenance. But even the funding was merely able to cover the initial payment, including the rental housing subsidy and the relocation fee. The Ludancun redevelopment project would cause the government a deficit of more than 1 million RMB. If the residents did not share the cost, how could the government initiate the demolition and construction? (Guo, 2004)

In July 12, 2001, without any notification or further opinion solicitation to the affected residents, the municipality announced the redevelopment plan and the compensation schemes to the residents. Three weeks later, the Housing Bureau engaged a temporary organization "Ludancun redevelopment office" on site to sign contracts with the affected residents. However, the compensation schemes triggered turbulent resistance from the residents. Some affected residents complained in an online forum:

> Those who determined the compensation scheme don't live in Ludancun, so they don't understand our concerns. Since our children all get educated in nearby schools, if we move far away from Ludancun, the access to their schools will be extremely inconvenient; if we rent a house close to Ludancun, the compensated rent is far from enough. Most importantly, we purchased houses of terrible quality from the government. Therefore, it is the government's obligation to finance and implement the redevelopment. Why should we share the cost of the redevelopment? The government should refer to the developer who built Ludancun for the reimburse. (SZhomeBBS, 2000)

Residents organized a property owner committee (POC) (as the application of aggregation rule) and promoted Chen Keren, a reputable professor, as the head. Through collecting residents' requirements, the POC as the representative of the community, submitted a petition letter to the municipal government (as the application of choice rule), in which they noted that:

> It is impossible for us, the property owners, to pay for the redevelopment project…In addition, the cost of the interior decoration should be compensated since many households spent a large amount of money on it. (SZhomeBBS, 2000)

The municipal government, however, did not attach much importance to the objections. In April 2002, the Housing Bureau promulgated a demolition permit as an official notice to the residents and forced the residents to sign an agreement with the engaged office. To pressurize the original residents, the Bureau set up a deadline for agenda setting—if the majority of residents did not agree with the scheme before April 28, 2004, the redevelopment would not be approved, and the residents would have to continuously suffer from the poor living environment. However, the communication between the government and the affected residents did not prove effective and smooth. The Housing Bureau lost its patience before the coming up of the deadline. In January 2003, based on its structural power, the Housing Bureau promulgated a public letter on site to force the property owners to relocate, which triggered turbulent opposition from the affected residents. The violent resistance resulted in the fierce conflict and finally excluded the officials from Ludancun (as an application of choice rule). An official of the Housing Bureau revealed:

> Some property owners rushed into the redevelopment office in Ludancun, and smashed windows and interior furniture. The officials were so scared and fled the scene…I admit that the opinion solicitation was not enough and the compensation scheme had not been fully discussed with the residents, which resulted in the conflict. (Zhou et al., 2014)

Facing the stiff opposition from the residents, the municipal government realized the requirement and satisfaction of the residents could not be ignored any longer. It thus proposed a new strategy —to purchase the property from the residents (as an application of pay-off rule) (Guo, 2004). This monetary compensation method was supposed to give the residents more options for resettlement. Though the compensation was rather higher than the households originally paid to their work units, it was not enough to purchase a new apartment with the same size in surrounding areas. The new strategy thus incurred immense oppositions from the community.

In June 2004, administrative reform in Shenzhen retarded the redevelopment process. The original Housing Bureau was arranged to combine with the land department to establish the Land and Housing Bureau. The combination resulted in the changes of managing authorities of Ludancun redevelopment and put the project into standstill. In the report to the Municipal government, the Land and Housing Bureau indicated that it was reluctant to support Ludancun redevelopment for two reasons (Zhou et al., 2014). Firstly, though the housing quality was poor in Ludancun, according to the Decrepit Housing Appraisal Standard, it did not meet the requirement to be demolished. Secondly, the financial infeasibility of the municipal government and the excessive resistance from the residents created massive obstacles to the implementation of the redevelopment. However, residents contended that the genuine reason for the project suspension was the absence of the official in the government to promote Ludancun redevelopment. *"The previous mayor was elevated to the provincial government so that no one in the local government supported and took over the redevelopment."* (Interview of HY, Jan-09-2017).

The deteriorating living environment caused mounting dissatisfaction of the residents. As the representative of the residents, the property owner committee wrote petition letters to the current mayor several times to propel redevelopment process. They also mobilized the residents to demonstrate with banners in the municipal government. With increasing petition from Ludancun, in 2007, the Land and Housing Bureau intended to reboot the Ludancun redevelopment and promulgated new compensation schemes. The residents would be compensated by either monetary indemnity or property right exchange. If the former was chosen, the affected residents would obtain monetary compensation which was 30% higher than the market price of their existing apartment. For the exchange of the property right, they would be compensated an in-situ apartment unit with the additional 5 square metres usable area. In addition, the local government promised that the reconstructed houses with the best location would be designated as resettlement housing. The cost of interior decoration, temporary rental housing and relocation would be reimbursed.

Since the new compensation schemes would put its financial feasibility in jeopardy, the Municipal government attempted to arrange a private developer to invest the project. To motivate the engagement of the private developer, the government planned to allocate a piece of land in the urban fringe free of charge to the operator as allowance. However, the worldwide financial crisis from 2007 to 2008 depressed the real estate market in China. As Shenzhen housing price dramatically decreased at that time (Lin, 2016), no private developer was willing to engage in Ludancun redevelopment. Meanwhile, the affected residents rejected the government's proposal and obstructed the involvement of the private developers (as an application of choice rule) because of the distrust to the market sector. The municipal government had to delay the redevelopment project again.

At that time, more than 80% of the original residents had moved out of the area and the remaining residents were still suffering from the deteriorating living environment. The vice chair of the property owner committee Ms. Wu indicated that,

> When Ludancun was built in the 1990s, it was a modern community. The residents were mostly well-educated and working as civil servants in the government and other public institutions. In 2009, most of the original households moved out and rented out the apartments at a relatively low price. The overall quality of the new tenants was quite low. Crimes mountingly occurred in the community, which led to a serious threat to community security and order. (Ma, 2009)

4.3.2 Equal Consultation Arena

The participatory redevelopment reform, as an exogenous influence, allowed the residents to join in the decision-making process, which substantially enhanced the interest protection of the original residents in urban redevelopment. The local government had to resort a more collaborative approach to obtain the residents' agreement for the accomplishment of urban redevelopment, which subsequently formed the equal consultation arena.

In 2009, to facilitate the construction of major infrastructure, Shenzhen municipality put forward a government-led redevelopment approach "Land preparation". As the pollution of the Shenzhen River had negatively influenced the living environment of the surrounding area for many years, Ludancun was chosen as a pilot to locate a sewage treatment plant by the Shenzhen River. To promote the project, the municipal government designated a deputy mayor to be responsible for the Ludancun redevelopment (as an authoritative resource) and also arranged specialized funding (as an allocative resource) to initiate the project as soon as possible. Upon the set of the pilot project, the municipal government arranged a structural safety and reliability test on the residential buildings in Ludancun. The result showed that the buildings were decrepit housing of Dsu Standard, which had to improve in a demolition and reconstruction way. On May 25, 2009, the municipal government formally determined to conduct the demolition and reconstruction of the Ludancun through the government-led approach and the Municipal House Development Centre (MHDC), under the administration of the Municipal House and Construction Bureau[3] (MHCB), was in charge. On the next day, the municipal government held a specified meeting to approve the Ludancun redevelopment agenda.

On August 7, 2009, the MHCB publicized the planning and design plan of Ludancun redevelopment through local newspapers and posters to solicit advice from the residents (as the application of information rule). With the collection of more than 200 comments from the property owners, the MHDC thus adjusted the plan. On December 11, 2009, the MHDC promulgated the redevelopment schemes including a compensation scheme, a resettlement scheme and detailed planning and design plan (as the application of pay-off rule). According to these three files, around 35% of the original residential area in Ludancun (26,000 square metres) would be expropriated by the local government for public green space with the sewage treatment facilities installing underneath. On the rest part of the land, 11 32-storey high-rise residential buildings would be built. Community supporting facilities, such as kindergarten, food market, community centre and postal offices would be provided while outdoor activity facilities would be constructed including basketball courts, tennis courts, gate-ball courts, outdoor fitness venues and other sports venues. The community would have an underground parking space with 1292 parking lots. After the redevelopment, the floor area ratio (FAR, of the construction land except the public park) would be increased to 4.53, which is 2.89-fold of the original one.

The local government informed the residents that according to the relevant regulations and the current redevelopment practice in Shenzhen, residents would not have to share the project cost. The relocation fee and rent occurred in the demolition and resettlement process would be subsidized according to the market price of the adjacent area. The Demolition Compensation Ratio (DCR) was 1:1 and owners could obtain an additional five square metres award because the addition of elevators in the high-rise residential buildings would occupy some usable areas. Owners were also allowed to purchase additional area for their new apartment in accordance with the market price of the adjacent apartment units. In order to attract the residents to sign agreements, the MHDC proposed an award mechanism (as a pay-off rule). After the issuance of the demolition notice, if the affected residents signed an agreement within 6 months, each household would be awarded an addition of 5 square metres for their new apartments; a reward of 30,000 RMB would be given if they signed an agreement within 3 months and then moved out within 1 month; if they signed agreement and moved out after three months but within six months, 10,000 RMB would be rewarded as bonus. The head of the MHDC noted that:

> Before the promulgation of the schemes, the MHDC put a lot of efforts into preparation. We held three public meetings with the affected residents to collect their opinions. We did not show the specific plan and schemes but did 'blow some wind' to the households. Then we consummated the plan and the compensation details based on the feedback. In order to create a good living environment for residents, the redevelopment schemes took the housing

[3] To further improve the institutional efficiency, Shenzhen Municipal government conducted an institutional reform to combine scattered public agencies into several major public departments in 2009 (Wang & Wu, 2009). The newly formed Municipal House and Construction Bureau took charge of the housing expropriation in Shenzhen.

orientation and the environmental resources into consideration. The relocation houses were orientating to south, the Natural Park in Hongkong whilst commercial and office buildings were located in the north. Compared with the compensation scheme in 2004, this scheme is far more beneficial to the affected residents. (Zhou, 2009)

Out of the concern about the resistance, the MHDC engaged a temporary agency in the residential area to answer questions and collect opinions from the residents. *"We are trying to avoid scary violence like what happened in 2003. This time, their resistance is eased."* (Zhou, 2009).

Despite the great efforts the government devoting lowered the tension, the majority of the property owners still were not satisfied with the compensation scheme. The POC collected the objections through interviews and online blogs and intended to feedback to the local government to bargain for fairer compensation terms (as the application of choice rule). The objections mainly include five perspectives:

At first, most residents perceived that the additional five square metres were not fair, since their expectation was raised by the higher DCR in developer-led residential redevelopment:

We are now living in a low-rise building. After the redevelopment, the addition of elevators and corridors in high-rise buildings will be counted in the area of our new apartment. How could five square metres be enough? We don't care about the five square meters award and will be satisfied as long as we could obtain the same usable area with our original apartment. In addition, private developers provide higher DCR in the developer-led residential redevelopment than in Ludancun. This is so unfair! (Residents comments in the Online blog, POC, 2009b)

Secondly, the plan that 30% of land would be freely required by the government triggered substantial oppositions. It would inevitably increase the development density of the new community and decrease the living quality (POC, 2009a). Thirdly, the property owners were carving for higher subsidy for the temporary rental housing as the proposed rental subsidy from the government was not enough to rent a similar house in the adjacent area (Interview of WDB, Dec-28-2016). Fourthly, the redevelopment only provided large apartments which were beyond the affordability of residents, especially who lived in small houses, because they had to spend a large amount of money to the extra area (POC, 2009a). Fifthly, the regulation that the house distribution in Ludancun was according to the cadre's position met with certain opposition. When Ludancun was built, due to the limited amounts of large houses, some cadres in high positions were allocated in un-matching and smaller units. Through the redevelopment, these households could exchange for larger units. However, the government set the deadline as April 28, 2002 for the cadres' identification, which caused opposition from some affected residents.

We were allocated a smaller apartment in 1990 according to my father's position in the municipal government. In 2004, my father got promotion, but he went to do business and left the government in 2007. Now, the government identified my father's position in 2002 rather than 2004 and we were not able to exchange for a larger house. This is so unreasonable! (Interview of YBX, Jan-17-2017)

To efficiently respond to the resident's requirement and direct Ludancun redevelopment, the municipal government intended to enlist more staffs from relevant departments to set up a specific organization to replace the temporary office from MHDC. However, after the administrative reform in 2009, each newly established department was struggling with its restructuring process, which resulted in extraordinary difficulties in the coordination among different departments. Just as the head of the LRMO noted that:

> The MHDC present to the mayor that more coordination among different departments is critical for Ludancun project. Otherwise, the redevelopment process would be still in a standstill. (Interview of WDB, Dec-28-2016)

To promote Ludancun redevelopment, the mayor deployed his power to coordinate the relevant departments. On April 22, 2010, the Ludancun Redevelopment Management Office (LRMO) was established with officials from several relevant departments engaging on site (as the application of position rule). The officials were optimistic about the redevelopment and claimed to the affected residents that after the approval of the demolition permit and the issuance of the demolition notice, the municipal government would start to sign contracts with the residents and initiated the demolition in the short term.

On January 19, 2011, however, a central policy from the State Council abolished the demolition permit scheme and encouraged more public participation in the residential redevelopment. As the Ludancun project did not obtain a demolition permit, it must operate in accordance with the new policy. The local government had to suspend Ludancun redevelopment and waiting for the establishment of the new redevelopment procedure.

The project stall caused considerable dissatisfaction of the residents. The POC kept pressurizing the government through the petition letters and mobilized actions (as the application of scope rule). Since the 2011 Summer Universiade (the World University Games) would be held in August in Shenzhen, the municipal government prioritized the maintenance of social stability. In July 2011, the MHCB posted a formal reply to the POC to appease the property owners. In the reply, the government indicated that the agenda setting and site selection of Ludancun redevelopment had completed and the formalized compensation and resettlement schemes would be issued in the short run. To pressurize the affected residents, the government claimed in the reply, *"…the successful redevelopment of Ludancun depends not only on the determination of the government but also on the support of all property owners."* (Shen, 2012).

In November 2011, to promote Ludancun redevelopment, the municipal government held several joint meetings to distribute obligations among the relevant departments. Since the formal institutional arrangement was not ready, the Municipal Government delegated the power of house expropriation to the MHCB and urged the LRMO to solicit opinions from the affected residents and amend the compensation schemes accordingly (Shen, 2012).

Once the decision was made, to expedite the scheme amendment, the grassroot government organization and self-organized committees joined in the group to persuade the residents to be cooperative. Officials in the LRMO specifically solicited opinions through face-to-face household survey. The work station, as a grassroot government organization with a closer social relationship with the original residents, assisted the LRMO to contact and negotiate with the residents. The POC and the senior resident association (SRA), as local self-organized institutions, voluntarily joined the group to explain policy details and smooth appeasement for preventing residents from petition (as the application of information rule). As the head of POC indicated,

> We have dedicated to the redevelopment throughout one decade so that we have a good relationship with the local officials who oversee the project. In the process of opinion solicitation, the POC, as a legally recognized community organization, played a dual role. On one hand, to remove illusions and unreasonable expectations from some residents, we set up an online blog and hold several public meetings to broadcast policy and share information. For instance, we claimed the donation of land to the government is necessary for any residential redevelopment in Shenzhen. On the other hand, we work as a representative to negotiate with the government and urge to publicize the amended compensation plan promptly. (Zhou et al., 2014)

However, the competition among the developers in the market-led residential redevelopment escalated the DCR even to 1:1.4 in June 2012 (Li, 2012). This news subsequently resulted in the immense dissatisfaction in Ludancun. The Ludancun redevelopment office urgently convened a meeting with the owner representatives and declared that the DCR 1:1.4 had not yet been approved by the government. To curb the malignant competition between the different developers and its negative influence on other redevelopment projects, the Shenzhen municipal government set an upper limit of the DCR as 1:1.3 in an internal meeting in 2012 (Interview of an official in Nanyuan Stree Office, Jan-12-2017).

The resident representatives thus endeavoured to relieve residents' grievance. As the deputy president of the SRA noted,

> We conveyed the idea to other residents that the redevelopment was aiming for the public interest of the whole community. The government was willing to assist us to improve the living condition but its consideration was not very careful. We, as residents, should adopt a synergic, not a confrontational attitude. With the cooperative attitude, we could require the government to adjust schemes according to our requirement, which would lead to a win-win situation. If we kept in a hostile attitude, the government might lose patience and then cancelled the redevelopment, resulting in a lose-lose situation. As most of the residents were not satisfied with the low DCR, we conducted several home visits to each household to explain the advantages of the government-led model and the shortcomings of the market-led model. All the efforts we committed took effects. (Interview of HY, Dec-09-2017)

After finishing further opinion solicitation, the office provided feedback to the Housing and Construction Bureau. The later proposed an amendment plan and submitted to the municipal government. Through several discussions at the municipal level, on November 20, 2012, the municipal government promulgated the amended compensation scheme to the public, in which responded a large portion of requirements (as the application of pay-off rule) (Zhang, 2013). The DCR was promoted to 1:1.16 while small apartment units were provided for households which were not wealthy enough. The subsidies for temporary settlement and the compensation for the interior decoration were also raised. To accelerate the redevelopment process, the municipal government listed the Ludancun redevelopment as a key project and required relevant departments to cooperate with it.

After the revision, the government conducted a survey about the support of the affected property owners. A total of 1144 written opinions were collected from the owners while 31 owners did not submit their advice. Among them, 1046 owners agreed (with more than 504 households submitting with comments), accounting for 89% of the total; 98 households did not agree, accounting for 8.4%. The left property owners could only verbally agree to be cooperative because they had to address some issues regarding approval of bank mortgage and issuance of house certificate. Based on the collected feedback, the Municipal government further added the currency compensation approach, increased the compensation for interior decoration and expanded the quote for additional purchases. The new compensation scheme subsequently obtained agreement from more than 90% households.

The institutional arrangement regulated that on completion of agreements from more than 90% residents, the government was able to issue an expropriation decision for the rest of uncooperative residents and initiate the property assembly process (Zhang, 2013). In August 2013, the municipal government approved the redevelopment schemes and delegated district governments[4] as the operator for the housing expropriation. The head of the LRMO noted, *"the policy issue of the municipal government has been completely resolved, leaving the implementation duty to the district governments."* (Interview of WDB, Dec-28-2016) On December 18, 2013, the Luohu and Futian District government issued the "Decision of House Expropriation" in Ludancun, which signified the end of the consultation period of redevelopment schemes.

Although the majority of residents had agreed to sign contracts with the government, the agreement signing was obstructed by some remaining issues. For example, a small group of residents had issues about mortgages, unclear ownership rights and outdated premises permit. The solving of these issues relied on the cooperation among several departments and institutions, such as Planning and Land Committee, property registration department, audit department and national banks. After the issuance of the house expropriation decision, to speed up the house expropriation, the mayor asked all departments in the municipal and district governments to fully cooperate with the MHCB. Under the authorization of the municipal government, the director of the MHCB held several meetings to coordinate the relevant departments and required the related departments and institutions to engage staff on site to open up a "green window" for these residents (as the application of position rule). This temporary window (in terms of an office) provided "one-stop" services for the residents, which greatly increased the efficiency of agreement signing process. For the

[4] Ludancun was located in both Luohu and Futian districts. So both district governments were authorized as land expropriators.

remaining uncooperative owners, the LRMO adopted a soft strategy (as the application of a choice rule) and arranged specialized officials to *"treat them with sincerity and convince them with reasons"*. (Interview of D, Jan-03-2017).

However, in order to accelerate the contract signing process, the government even adopted illegal means in the face of the ambiguous property rights (as the application of choice rule). According to the interview of the head of SRA, *"after the death of a veteran cadre, the house was left to his second wife and the son. The son, born by the cadre's first wife who was passed away a long time ago, perceived that the unit in Ludancun belonged to his original patients. As both of them were dead, he should inherit it. The second wife, who was living in Ludancun, wanted her own 50% of property right and therefore refused to relocate. According to the law, the property right should be equally distributed to the son and the second wife, and only both of them signed the contract could the house expropriation be implemented. However, the LRMO simply obtained the agreement of the old lady and eventually demolished the house without the agreement of the son. This action violated the law but significantly reduced the difficulty of property assembly. If the son sued the municipal government in the future, he would only be compensated half of the estimated price of the demolished house, which is much easier than the efforts to deal with the controversy between the stepmother and the son. The government dares to do illegal things based on its administrative power."*(Interview of HY, Jan-09-2017).

Through the diverse strategies, since the formal agreement signing started on December 21, 2013, more than 96% of the property owners had signed the contract within four months. Upon completion of the agreement from the majority of the households, on the one hand, the MHCB hired a housing demolition company through open bidding to initiate the demolition; on the other hand, the officials kept communicating and negotiating with the last 20 uncooperative households. According to one of these households, A'Bao,

> We were against the regulation of housing exchange and struggled for a larger apartment which was matching to my father's position before he left the government in 2004. To persuade my family to be cooperative, the officials kept calling my parents for gentle and kindly negotiation. Finally, my family and the MHCB came to a compromise that we could obtain an apartment with larger area but it had to be in Shenyuncun, a public housing area of the municipal government because the location and value of the apartments in Shenyuncun is not as good as Ludancun. We eventually accepted the offer in January 2014. At that time, the buildings with all households moving out had been demolished, with only several buildings standing as around 20 households were still negotiating with the MHCB. (Interview of YBX, 2017)

Through the efforts of the MHCB, another ten households eventually signed the agreement, leaving last ten households which did not agree with the compensation scheme and opposed to the house expropriation. *"There were two households without property deeds. The reason why they were able to live in Ludancun because of their good relationship with the previous director of the Housing Bureau. Without property ownership, they were willing to obtain new apartment units through the redevelopment...For other uncooperative households, they were bargaining for higher compensation by deliberate refusion and violent resistance."* (Interview of HY, 2017).

The land assembly was trapped in a dilemma as the last households possessed a tough attitude towards the relocation. The negotiation strategy would not be effective to persuade the remaining uncooperative residents.

4.3.3 Mandatory Enforcement Arena

To address the dilemma, the Municipal government began to resort to some mandatory strategies to facilitate the compliment of the residential redevelopment, which structured the mandatory enforcement arena. As Ludancun redevelopment was conducted by the local government and aimed at public interest, the local government adopted the compulsory acquisition strategy to withdraw the property right from the uncooperative residents (as the application of scope rule). In February, 2014, the LRMO formed a joint team with the work station, the Public Security Bureau, the Police Station, the Court and the local media. The team broke into the houses of the remaining uncooperative households, registered and listed all the items and shipped them to the public housing which the local government prepared for the households. These uncooperative residents without property right had the right to live in the public housing but they had to continuously pay rent. The households who had property ownership but refused to sign agreement had the right to choose between monetary or in-kind compensation, according to the existing compensation scheme. To make the process open and transparent, the public media was recording the whole process by video. The municipal government claimed the process as "harmonious demolition". On completion of the housing expropriation, the demolition company cleared the existing constructions.

Once the demolition was accomplished, the land was withdrawn by the municipality and transferred to the local land bank. On December 31, 2014, the municipal government promulgated the land use auction notice about Ludancun (as the application of boundary and pay-off rules). Different from the general land auction which bids for a higher land price, the land price and the amount of public and relocation housing had been fixed in this project (Southern Metropolis Daily, 2015). Developers were required to bid downwards for the marketable and salable housing area. According to the interview of the head of the Land and Real Estate Trading Centre, there were three reasons for the innovative land transfer approach (Interview of LCS, Dec-01-2016). At first, it was able to decrease the negative impact of excessively high development density and reduce the accordingly dissatisfaction of the original property owners. Secondly, the extremely high land transfer fee might generate an impression that the municipal government sought immoderate profit, which might hurt the reputation of the land preparation approach. Thirdly, as the excessive profit for the government was constrained, so was reduced the residents' expectation and speculation for the succeeding projects.

Besides, as the first successful land preparation case, the municipal government intended to set Ludancun redevelopment as a remarkable and exemplary project for the following practices so that the requirement of construction technique and efficiency was extraordinarily high. It required the developer to utilize sustainable technology for construction waste treatment and carbon emissions reduction in the construction process and to carry out housing industrialization technical for commercial housing. All the new residential buildings should be verified as "green buildings".

> These initial restrictions on one side were able to promote the technological progress of the residential construction in Shenzhen and brought extensive social benefit; on the other side, through the initial restrictions, inexperienced and incompetent developers would be excluded." (Interview of LCS, Dec-01-2016)

After more than 60 rounds of competition, the China Overseas Land and Investment LTD (COLI hereafter), a central-state-owned enterprise with abundant financial and technological strength as allocative resources, won the bid. Through waves of business merger in the past two years, COLI received RMB 33.8 billion capital injection and 10.92 million square metres of land reserves from the China State Construction Engineering Corporation (Liu, 2015). The merger of the housing industrialization leading company– Hailong component plant,[5] also empowered the COLI to stand out among the competitors.

The competition process considerably reduced the profit margin of the developer. The housing for public and relocation was 144,000 square metres with only about 39,000 square metres commercial housing for the COLI (Wu, 2015). To improve the commercial value of the buildings, the developer meticulously utilized advanced materials for construction and equipped the interior spaces with luxury decoration and furniture. Because of these strategies, the new buildings were transformed into high-end commercial apartments with the price to 150 thousand RMB per square metre, the highest ever apartment price in Shenzhen (as the application of pay-off rule) (Wu, 2015). On January 23, 2018, on completion of the reconstruction, the original residents moved in the newly constructed apartments, which signified the accomplishment of Ludancun redevelopment.

4.4 Exogenous Influences and Their Impacts on Power Relations

The exogenous influences that had significant impacts on power relations in Ludancun redevelopment were composed of three categories: the central policies, the local policies and attribute of the community. The central policies served to impact on power relations from four perspectives: housing reform, property right reform and

[5] Hailong has been experiencing housing industrialization in Hong Kong for more than 20 years, and its construction level has passed the British quality management standard which is considered the highest technical requirement in the world.

the abolition of demolition permit institution. The housing reform transformed public housing from the asset-holding work units into privatized properties and switched the original tenants to the de facto property owners, generating the awareness of private property right of the affected residents. This awareness presented the affected residents' significant structural power resources and empowered them to take mobilized actions to resist unfair compensation terms in the redevelopment process in 2003. The property right reform further formally legitimatized the protection of private property rights and granted substantial structural resources to the residents to oppose improper treatment in the redevelopment cases. The participatory redevelopment reform enhanced the importance of urban redevelopment and pressurized the local government to share interests with the affected residents to facilitate redevelopment implementation. It triggered the emergence of the new government-led redevelopment mechanism with public participation arrangement at the local level, which provided structural power resources with the residents to join in the decision-making process. The abolition of demolition permit institution in 2011 restrained the deployment of the state power and further formally institutionalized the public participation mechanism, which granted substantial structural power resources to the affected residents to influence the redevelopment process.

Local policies have a great impacted on power relations. Once the land preparation approach initiated in 2009, the municipal government arranged specialized funds to support the experiment cases, which strengthened the allocative resources of the local government to facilitate the redevelopment. The policy also authorized the local government structural power to initiate a government-led redevelopment project and established a procedure to ensure the participation of the affected residents as structural resources. In line with the central policy in 2011, Shenzhen promulgated its housing expropriation policy in 2013, in which public participation mechanism was institutionalized in the government-led residential redevelopment process in Shenzhen. The affected residents obtained critical structural power to negotiate with the local government and participate the decision-making process.

The attribute of the community impacted on the local power relations includes the physical attribute and social composition and values in the community. The premium location and large size of the community signified great return of Ludancun redevelopment, which formed strong allocative resources for the affected households to negotiate with the local government when bargaining compensation terms. The inferior living conditions and poor building quality of the government-built community added up to the urgency of Ludancun redevelopment, which can be considered as allocative resources to the affected residents to influence the government's decision about redevelopment. Some property owners, as previous senior cadres in the local government, possessed strong authoritative resources to affect government's decision regarding their high-ranked cadre status, reputation and personal influences. Meanwhile, because of their working experience and social attributes, the vast majority of the residents in Ludancun held a firm belief on the local government and refused to involve the private developer to conduct redevelopment (as the application of scope role). The above factors managed to result in the application of the government-led approach in Ludancun. In addition, strong social networks were formed among the affected residents since most of them have lived together for three decades, which granted them strong authoritative resources to form self-organized associations and facilitate redevelopment.

4.5 Endogenous Influences and Their Impacts on Power Relations

The interactions between the three main stakeholders—the local government, the affected residents and the developer—resulted in the production and reproduction of power relations in Ludancun redevelopment through the continuous configuration of resources and rules. The local government originated its power in three categories of power resources. Structural resources mainly included its ruling position, land ownership, definition of regulations and policies and jurisdictions on land use and development intensity. To supplement its structural resources, it also obtained its power from capital investment as the allocative resource and social relationships as the authoritative resource. In the first arena, the government deemed itself had a higher position than the affected residents, a traditional powerless stakeholder. Without ample information exchange with the residents, it simply provided unfair compensation terms and then intended to adopt a forceful approach to implement the redevelopment in the short term (as the application of choice rule). However, the rule triggered fierce opposition from the affected residents. Based on their high social status and private housing ownership, their mobilization hindered the promotion of the redevelopment. To pacify social resistance, the local government thus shifted its strategy towards a monetary compensation approach (as the application of pay-off rule) to increase compensation. The compensation terms appeared to be unfair and unsatisfied to the latter since the local government did not invite the affected residents into the decision-making process. The new rule failed to take effect reflected by the failure in obtaining supports from the residents. In the second arena, the participatory redevelopment reform significantly empowered the affected residents and constrained the structural power resources of the local government. The local government subsequently changed its position from dominant to equal and set up a specialized office as a local agency to negotiate with the affected residents (as the application of position rule and information rule). Through public media, the government promulgated redevelopment agenda, project plan and compensation terms to the public for opinion solicitation, ensuring the decision-making process as fair and transparent (as the application of information rule). Based on the collected feedback from the residents, the local government escalated the compensation schemes to share more interests with the affected residents (as the application of pay-off rule). The frequent and intensive application of these rules came into force as the local government succeeded in obtaining supports from the vast majority of the affected residents to accomplish the redevelopment schemes setting. To simplify the relevant procedure of property assembly, the government set up an integrated office on site to coordinate all the property-related departments (as the application of position rule). Furthermore, it also adopted some illegal actions (as the application of choice rule) to smooth the way of agreement signing (as the application of choice rule). These rules served to facilitate the property assembly which was reflected by the rapid agreement signing process. In the third arena, the government began to adopt mandatory strategies to finalize the property assembly and construction. The local government formed a joint team to conduct housing expropriation from the last uncooperative residents (as the application of scope rule). On completion of that, in land auction process, to prevent extremely high development density, the local government executed its administrative power to fix the FAR and establish detailed requirement in selecting developers (as the application of boundary and pay-off rules). The rules took critical effect in facilitating the property assembly and reconstruction process.

The affected residents in Ludancun are in possession of abundant power resources in terms of structural and agent perspectives. They obtain power from the structural resources including the property ownership and the public participation mechanism of the participatory redevelopment policy; meanwhile, both allocation resources (e.g. premium location, large community size and poor living environment) and authoritative resources (e.g. close community relationship, reputation and personal influences on current cadres) significantly empower them to make a difference in the decision-making process. The de facto property right and their social status of senior cadres empowered them to take mobilized actions to effectively resist the unfair compensation from the local government (as the application of aggregation and choice rules) in the first arena. After the participatory reform, residents obtained strong structural resources to participate in the decision-making process. The well-maintained social relationship and residents' higher level of education both contributed to the formation of the self-organized associations, such as POC and SRA. These associations played an important role to associate the local government with the affected residents in the redevelopment process. Firstly, they assumed the resident representative to collect residents' comments, submit to the local government (as the application of information rule), organize mobilized activities to pressurize the local government (as the application of scope rule) and negotiate with the local government for fair compensation (as the application of choice rule). With high frequency, these rules came into force to unite the affected residents and obtain fair compensation terms when bargaining with the local government. On completion of the compensation terms, they transformed to be a power broker of the local government to elaborate the redevelopment plan details and promulgate the advantage of the government-led mechanism with the affected residents (as the application of information rule). These rules eventually facilitated the agreement signing of the vast majority of the affected residents.

The developer works as a contractor based on its allocative resources, including expertise and financial strength. To increase its profit margin, the developer built Ludancun into a luxury community with the highest apartment price in Shenzhen (as the application of pay-off rule). However, without structural power resource, the rule of the developer failed to meet its objective.

4.6 Outcome and Evaluation

4.6.1 *Outcome*

Ludancun redevelopment is the first successful and the only government-led dilapidated residential redevelopment project in Shenzhen. It went through a journey of more than 18 years, with 9 years of consultation for redevelopment agenda setting, 5 years of negotiation for redevelopment schemes, 1 year of property assembly (more than 1175 households signed the agreement within 8 months, and housing expropriation and demolition only took 82 days) and 3 years of construction.

After the redevelopment, the FAR has been escalated to 4.53, almost 3 times of the original FAR, signifying the immense increase of the development density. Around 35% of the land was expropriated by the municipal government for constructing a public park with a sewage treatment plant underneath. The 24 seven-storey buildings were demolished, and 11 high-rise residential buildings were built (one 32-storey, two 36-storey, two 37-storey, four 39-storey, one 44-storey and one 46-storey buildings). The residential buildings which are adjacent to the Shenzhen River with premium view were distributed to the original residents, while the other two residential buildings with inferior landscape values are arranged for the developer for commercial sale. Provided were community supporting facilities, such as one kindergarten, one food market, one community centre, one postal office, more than 1200 underground parking lots and outdoor activity facilities including basketball courts, tennis courts, gate-ball courts, outdoor fitness venues and other sports venues. With comprehensive facilities, Ludancun transforms from a dilapidated residential area with the most terrible living condition into the most modern and luxury community in Shenzhen.

In addition, the on-site relocation rate of residents is beyond 98%, which is extremely high compared with the previous state-led redevelopment cases (e.g. Shin, 2009). Except for two households who choose offsite relocation and ten households who refused to relocate, others with formal ownership all relocate to the newly constructed on-site community.

4.6.2 *Equity Evaluation of the Outcome*

The local government devoted considerable efforts and invested huge resources in the government-led redevelopment. The manpower and administrative costs were supposed to be substantial for the negotiation of redevelopment agenda and compensation schemes and the implementation of a property assembly. The funding granted from the local government to initiate the building quality test and land clearance was assumed to be sizeable. Therefore, the government is expecting high gains in this process with land use fee as 888 million RMB and 35% of the original land was expropriated at few costs for the construction of urban infrastructure. The substantially improved urban environment generated social benefit to the government as well as the public.

Residents purchased the public owned housing from their work units at a low price in the 1990s. After the redevelopment, because of the 1:1.6 DCR and the additional five square metres award, the cooperative residents gained a newly constructed apartment with larger size than the original apartment. Residents were also permitted to purchase extra areas for their new apartments at a market price. The cadres who were in high positions but gained unsuitable smaller apartments previously were allocated larger apartments matching to their positions. With newly constructed and comprehensive facilities in the community, their living condition was substantially improved, and the value of their property was significantly escalated. For property owners who refused to sign the agreement, the local government provided either newly constructed apartments or monetary compensation in accordance with the size of their original properties. Even the "uncooperative tenants" without legal property rights were reasonably treated and they were relocated by the local government to public housing with only paying relatively low rent.

The developer paid the land-based taxes and took charge of the reconstruction of the new community. Among the 11 newly constructed residential buildings, only two were allocated to the developer as commodities to cover the cost of reconstruction. To increase its profit, the COLI endeavoured to provide high-end community facilities and build luxury commercial apartments in the newly established residential area, which in turn significantly benefited the original residents and raised the price of the new relocation apartment. However, in 2016, to control the rocketing housing price and limit speculation in real estate market, the municipal government set a roof price for the sale of commercial apartment (Interview of the staffs in COLI, Dec-08-2016). This action substantially influenced the sale of the apartment and destroyed the developer's profit. According to the interview, most luxury apartments in Ludancun were reserved by the developer who is waiting for the cancellation of the price limitation (Interview of the staffs in COLI, Dec-08-2016). In this case, the interest of the developer was not appropriately protected.

4.6.3 Institutional Adaptation of the New Power Structure

Stakeholders' adaptability to the structural changes was differentiated into different levels in Ludancun. The local government has long been criticized for excluding the affected residents from decision-making process. Since the authoritative management arena was framed in 2001, the requirement of the affected residents had been ignored. Without solicitating support from the majority of residents, the local government attempted to enforce the residents into a contract process. The fierce strategy caused violent conflicts between both parties. To maintain social stability, the government shifted to some market-oriented approaches to facilitate the redevelopment. However, all of these measures did not aim to address the requirement of the affected residents but merely to avoid excessive financial burden of the local government. With the long-term preservation of authoritative management, the local government manifested its path dependence of socialist governance.

Due to the participatory redevelopment reform, the residents were empowered to participate in the decision-making process and to share interests generated from the redevelopment project. The municipal government swiftly adapted to the new power structure. By waiving the coercion strategy, the government transformed to be collaborative. The government arranged specialized funding and a vice mayor to coordinate relationship among different governmental departments and engaged one specialized office on site to communicate and negotiate with residents. Based on the collected feedback, the government escalated the compensation to share more interests towards the affected residents. Once the higher DCR in the market-led redevelopment projects triggered immense grievance of the local residents, the municipal government set a pay-off rule to not only set an upper limit for the DCR, but also escalate the DCR in the local project. These strategies eliminated the dissatisfaction and curbed the speculation behaviours.

To deal with difficulties in the property assembly process, the municipal government adopted a compulsory strategy to tackle the last residents, on completion of the agreement from the majority. When auctioning the land, the government adopted a regulatory strategy to fix the land price and required the developer to bid for lower development density. These mandatory strategies facilitated the implementation of Ludancun redevelopment.

The residents made a quick adaptation to the institutional change in Ludancun. Once the municipal government firstly published the compensation scheme in 2001, without a legal channel to express their requirement and participate in the decision-making process, some residents damaged the on-site office and expelled the staff in the government to resist the decision of the government. Through the fierce strategy, these residents excluded the official and revealed their determination to protect their interests. When the institutional reform delayed the process in 2004, the residents adopted a less fierce strategy—they appealed to the government through public letters, petition and demonstration under the organization of the POC. After 2009, the new institutional arrangement set up a formal procedure for residents to participate the decision-making process. The POC thus formed a legal channel with the engaged redevelopment office to express residents' requirement. Affected residents then replaced the antagonist attitude with a cooperative attitude to request fair treatment and compensation scheme.

4.6.4 *Transaction Cost and Project Efficiency*

In the Ludancun project, transaction costs incurred for setting redevelopment agenda and compensation terms between the local government and the affected residents. Field evidences show that transaction costs in Ludancun are associated with the local government's behaviours and heterogenous interest of the affected residents.

The authoritative management arena was characterized by power asymmetry. The local government was authorized significant power resources to dominate the redevelopment process while the affected residents have few power resources to participate the decision-making process. The powerful side deemed its position as higher and began to adopt forceful strategies to compel the existing households. This resulted in severe resistance and project delay and significant hold-up cost was incurred. To address, the local government intended to resort more market power to promote the redevelopment. However, the local government did not reposition itself to an equal position, which was reflected by constant ignorance of residents' requirement, the agenda setting was prolonged to add up transaction cost.

The heterogeneous interests of the affected residents protracted the negotiation process between the residents and the local government, decreasing the process efficiency. After the participatory redevelopment reform, the residents were empowered to join the decision-making process and the local government shifted to an equal position for negotiation with the former. With more than 1200 households involved in the bargaining process, the size of the involved stakeholders in decision-making was considerably increased. Different households had varying and diversified requirements which the local government had to make efforts to address and the negotiation process was extended, escalating the transaction cost.

4.7 Discussion

This section discusses the advantages and limitations of government-led power relations in delivering an urban redevelopment project. The successful implementation of the government-led case demonstrates its effectiveness in facilitating residential redevelopment. The retention of the government's authority for compulsory acquisition helps maintain a delicate balance of power between the local government and affected residents, thereby contributing to the completion of housing expropriation from the remaining uncooperative households. Moreover, the involvement of the municipal government enhances coordination among relevant local departments, substantially accelerating the process of property assembly.

As a government-led project, the Ludancun redevelopment benefited from the allocation of significant administrative resources, including the direct involvement of a specialized vice mayor and its designation as a "key project". During the land auction process, the government was able to exercise its dominant authority to adopt regulatory strategies that prevented excessively high development intensity and avoided potential negative externalities for inner-city infrastructure. In addition, the institutional arrangement of the government-led approach prohibited developers from participating in consultation and property assembly phases, restricting their role primarily to construction activities. By postponing developer involvement, this arrangement impeded the formation of a growth coalition between the government and developers, thereby enabling residents to capture a larger share of project benefits. Taken together, these institutional and power-related factors contributed to the successful delivery of the government-led redevelopment.

The evident drawbacks of the government-led approach also pose obstacles to its replication in other redevelopment projects. This approach struggles to manage the number of involved households and, consequently, the diverse demands and heterogeneous interests, which significantly prolongs the negotiation process over compensation schemes between the government and residents. In the absence of a substantial market operator, as is present in developer-led programs, multiple governmental agencies must be directly involved, creating considerable challenges for horizontal coordination within the local government. Moreover, the adaptability of the government-led mechanism is limited.

The approach is also unsustainable and difficult to generalize to typical residential redevelopment projects, as it requires substantial administrative resources and financial support from the local government. Furthermore, the strong-state authority inherent in this model can have adverse consequences. To expedite redevelopment, the local government at times adopted opportunistic measures that undermined the legally protected interests of certain residents. Simultaneously, the strategies employed to control development density compromised the financial feasibility of projects, further threatening the long-term sustainability of the government-led program. Overall, the retention of strong-state power, while facilitating project implementation, presents a latent risk of infringing upon the interests of other stakeholders.

Under the arrangement of the local government, which retained dominant authority, each actor in the government-led case derived certain benefits. However, this consensus was primarily achieved through the redistribution of land value increments among the involved stakeholders. It would be misleading to suggest that the government's role has reverted to that of a welfare provider, as in the socialist period. Rather, this case supports the claim of Wu (2018) that consensus for urban change is reached through technical rationality rather than a need to redistribute welfare.

4.8 Summary

This chapter examines the power relations in the government-led participatory redevelopment process, as summarized in Table 4.1. The local government holds dominant authority and oversees the entire decision-making process, including project approval, the formulation of redevelopment schemes, property assembly and the selection of private developers. At the same time, affected residents are substantially empowered to play a meaningful role in negotiating fair and satisfactory compensation with the local government. In contrast, the involvement of private developers is delayed, restricting them primarily to a contractor role with minimal influence.

Table 4.1 Summary of power relations in the government-led case

Actor	Rules employed	Resources mobilized
Phase 1: Authoritative management (2000–2004)		
Local government (housing bureau, etc.)	• choice rule • information rule • pay-off rule • position rule	• structural resources • authoritative resources
Residents (primarily retired cadres)	• position rule • aggregation rule • choice rule	• authoritative resources
Equity: Low Adaptability: Low Transaction costs: High		
Phase 2: Equal consultation (2009–2013)		
Local government (MHDC, LRMO, district agencies)	• information rule • pay-off rule • position rule • scope rule	• structural resources • authoritative resources • allocative resources
Residents (POC, SRA representatives)	• information rule • scope rule • position rule	• structural resources • authoritative resources
Equity: Improved Adaptability: High Transaction costs: Moderate-to-high		
Phase 3: Mandatory enforcement (2014–2018)		
Local government (LRMO, courts, police)	• scope rule • Boundary & pay-off rule • choice rule	• structural resources • authoritative resources • allocative resources
Developer (COLI)	• pay-off rule • boundary rule	• allocative resources
Remaining residents (holdouts)	• position rule • information rule	• structural resources
Equity: Mixed Adaptability: Moderate Transaction costs: Low		

Consequently, power relations between the local government and affected residents shift from conflictual to collaborative. This inclusive and cooperative dynamic facilitates the successful implementation of the redevelopment project.

References

Feng, J. (2000, Novermber 24). The government tended to invest 700 millions to redevelop Ludancun. *Yangchang Evening News*.

Guo, Y. (2004, May 16). The difficulties of Ludancun redevelopment. *The Economic Observer*. http://finance.sina.com.cn/g/20040516/1832762512.shtml

Li, Z. (2012, June, 21). The highest demolition compensation ratio in Shenzhen --1:1.4. *Southern Metropolis Daily*. http://house.ifeng.com/detail/2012_06_21/25467700_0.shtml

Lin, L. (2016). Shenzhen housing prices plummeted in 2008 triggered blossomed owners opposition. *Phoenix Real Estate*. http://sz.house.ifeng.com/detail/2016_03_10/50716925_0.shtml

Liu, J. (2015). After redevelopment, the dilapidated residential area of Ludancun turns into the most expensive apartment in Shenzhen. *Time weekly*. http://www.kaixian.tv/gd/2015/1110/727318_2.html

Ma, X. (2009). The project delay of Ludancun redevelopment should be prevented. *Jing Newspaper*.

POC. (2009a). *Residents concerns about the Ludancun redevelopment*. http://blog.sina.com.cn/s/blog_636f52210100g3rg.html

POC. (2009b). Suggestions for the Ludancun redevelopment plan. http://blog.sina.com.cn/s/blog_636f52210100g3r4.html

Shen, W. (2012, January 7). Ludancun redevelopment has been delayed for 10 years. *Shenzhen Evening News*. http://news.sznews.com/content/2012-01/07/content_6372420.htm

Shin, H. B. (2009). Residential redevelopment and the entrepreneurial local state: The implications of Beijing's shifting emphasis on urban redevelopment policies. *Urban Studies, 46*(13), 2815–2839.

Southern Metropolis Daily. (2015). The land auction of Ludancun will take place in February 6, which may trigger intensive competition. *Southern Metropolis Daily*. https://xw.qq.com/cmsid/20150122010003800

SZhomeBBS. (2000, August 11). *The Ludancun redevelopment will commence soon*. http://bbs.szhome.com/30-30-detail-882.html

Wang, C., & Wu, J. (2009). Newly constructed department in Shenzhen is listed as a key step in the reform of the major department system. *Chinese Government Network*. http://www.gov.cn/jrzg/2009-09/08/content_1411981.htm

Wu, F. (2018). Planning centrality, market instruments: Governing Chinese urban transformation under state entrepreneurialism. *Urban Studies, 55*(7), 1383–1399. https://doi.org/10.1177/0042098017721828

Wu, J. (2015, December, 4). The price of Ludancun apartment is 15 thousand per square metre. *Real Estate Opinion Network*. http://www.guandian.cn/article/20151204/168484.html

Zhang, W. (2013). The Ludancun redevelopment has been accerlerated and the new compensation scheme is published. *Nanfang Daily*. http://epaper.southcn.com/nfdaily/html/2013-01/25/content_7162677.htm

Zhou, C. (2009, December, 18). The Ludancun redevelopment needs wisdom from the local government. *Nandu Daily*. https://news.sina.com.cn/c/2009-12-18/042516791405s.shtml

Zhou, C., He, Y., Guo, Q., & Chen, W. (2014, May 29). *The historical records of Ludancun*. http://epaper.oeeee.com/epaper/H/html/2014-05/29/content_3248972.htm?div=-1

Chapter 5
Developer-Led Participatory Redevelopment in Mutoulong

5.1 Introduction

This chapter investigates the power relations among the main stakeholders—the developer and the affected residents—in a developer-led redevelopment project. Section 5.2 provides the background of the case. Section 5.3 examines the social practices and evolving power relations throughout the redevelopment process. Section 5.4 analyses the impacts of exogenous influences, while Sect. 5.5 considers the effects of endogenous influences. Section 5.6 presents the spatial and social outcomes of the project and evaluates the associated power relations. Section 5.7 concludes the chapter.

5.2 Case Context and Community Attribute

Mutoulong is a large Dilapidated Residential Area (DRA) located at the central Luohu District, as shown in Fig. 5.1. The area covers 100,000 square metres, with 57 seven-storey buildings (1340 apartments), built by the local government in 1983. The original FAR was 1.55. Its original occupants were workers and staffs in major state-owned-enterprises (SOE, e.g. Shenzhen power supply company) and public institutions (e.g. Shenzhen customs). Through the 1990s housing reform, these apartment units were privatized and sold by asset-holding work units and become market transferrable. The average age of the heads of these households is above 60 and some are laid-off workers through the previous SOE reforms in the 2000s. Since the majority of residents in Mutoulong have lived there for over 30 years, a strong social network exists in the local community. However, differ from collectively governed "urban villages", residents in Mutoulong are of a loose association. There is not any self-organized association in the community.

X. Li, *Participatory Redevelopment as Contested Arenas in China*,
https://doi.org/10.1007/978-981-95-9394-1_5

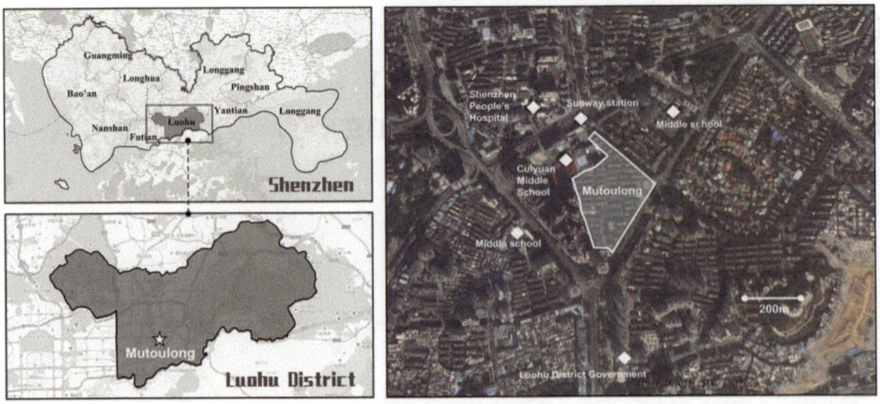

Fig. 5.1 The location of Mutoulong, Shenzhen

The public facilities of Mutoulong are very convenient with one subway station, one public large hospital, one reputable primary school and one prestigious middle school easily accessible with 200 m. With only 5 min' walk to the CBD of Luohu District, Mutoulong occupied a premium location. The original resident noted that the community, as a dilapidated but thriving residential area, once had a peaceful and liveable living environment (Interview of two residents, Dec-07–2016). Residents used to come out to talk to each other after dinner and enjoyed the cooling environment under the towering trees. Over 30 years passed, building conditions have seriously deteriorated: the walls were cracked and leaking, and fire emergency equipment was relatively backward. Heavy downpour always resulted in flooding in the community as its ground is 40 cm below the main road in Shenzhen. The sound location of Mutoulong was contradictory to its low density and poor living condition. Given the poor living quality, most residents are supportive for the redevelopment.

5.3 Power Arena Analysis in Ludancun Redevelopment Project

The Mutoulong redevelopment adopted a developer-led approach. Since the project commenced in 2009, the developer, Yitian Group CO. LTD. (Yitan hereafter), was granted power to trail 3 strategies to secure residents' support and consent. These strategies were implemented through three actions sequenced in time and unfolded in three "relational power arenas", namely the publicity, physical and negotiation arenas. This allows the analysis to focus on owner-developer power interactions.

5.3.1 Publicity Arena

At the project commencement, Yitian adopted a "publicity strategy" to obtain residents' documented supports. When it first was attracted by Mutoulong redevelopment in 2007,[1] Yitian confronted the competition of the Urban Construction Development Company (UCDC hereafter). UCDC, as the original property management service provider on site, had developed a close social relationship with the original residents. It initially offered a DCR as 1:1.15 and obtained almost 30% of the supports in half a year. To compete with the UCDC, Yitian employed two former local government senior officials with attractive remuneration to promote its public image. As a result, increasing residents accepted that Yitian is a reputable developer with strong financial capacity for urban redevelopment projects. Yitian then offered its DCR as 1:1.2 and promised a more attractive compensation and relocation scheme to the affected households (as the application of pay-off rule). The competition between two companies escalated the DCR from 1:1.2 to 1:1.25 in the short term. In 2009, UCDC announced a DCR as 1:1.26 and claimed it as the highest ever DCR in Shenzhen, but Yitian shortly raised it to 1:1.3. Considering that the continuing increase of DCR would eventually diminish the profit of both companies, Yitian and UCDC came to terms (Interview of LPL, Nov-29–2016). UCDC agreed to quit the competition and Yitian began to take charge of the redevelopment project. As the previous property service provider quit, Yitian promised to provide property management service for the residents until the redevelopment has been accomplished.

As the single engaged developer on site, Yitian began to directly negotiate with the affected residents regarding their expectation of the redevelopment and requirement of compensation (as the application of information rule). It held several public hearing meetings with the majority of the affected residents and then conducted several rounds of negotiation with a small group of resident representatives. On completion of that, Yitian determined the DCR as 1:1.3 accompany with rental subsidy and compensation for relocation and interior decoration (as the application of pay-off rule). This offer was recognized by the majority of residents. As the cooperative residents noted,

> At that time, 1:1.3 was the highest DCR in Shenzhen. Although the new community will become a community with high-rise apartment complex, the area of each unit will increase by 30% (according to the DCR), which can offset the shared area of the elevators and corridors…The news reported that the Mutoulong redevelopment would not generate huge profit for Yitian since the DCR was so high. But the location and surrounding public facilities of Mutoulong are superb, and the current density is very low. Even though the DCR is 1:1.3, Yitian still have a considerable margin to seek profit since the housing price in Shenzhen is rocketing. (Interview of two residents, Dec-07–2016).

In 2010, since Yitian obtained the supporting documents from over 70% (thus meeting the two-third threshold) of the households, the district government approved the redevelopment agenda and awarded Yitian as the formal candidate for Mutoulong

[1] The local government promulgated draft policy for public opinion solicitation in 2007. Some developers who perceived residential redevelopment projects as an excellent profit-seeking option entered dilapidated residential areas in that year.

redevelopment (as the application of position rule). Yitian consequently commenced signing contracts with the affected residents.

On the completion of the agenda, to accelerate the contract signing, Yitian promised to share more profit with the residents. By late 2010, Yitian announced in a press conference that it had secured 10 billion from the Industrial and Commercial Bank of China, incentivizing it to increase the compensation to attract more cooperative residents. The manager then promised to the attending residents that "*…[w]e have offered the highest ever Demolition Compensation Rate (DCR 1:1.3)*[2] *in Shenzhen to benefit the affected residents. In addition, we will reimburse the relocation grant to cover moving expenses and pay an allowance to cover rental expense until the affected residents move back to their new apartments. As more than 1000 residents have signed the agreement, we predict that about 90% (about 1200) of the residents will sign contracts before April, 2011. Upon completion of 90%, we will apply for the demolition permit to implement the demolition and construction in May, 2011*" (Lu, 2010). The successful negotiation of a 3 billion RMB loan with the China Construction Bank allowed Yitian to further increase the attractiveness of its compensation plan (as the application of pay-off rule). Residents who relocate within one month after signing the contract would receive an extra 30,000 RMB. If all owners in the same building consented, each would be rewarded with extra 50,000 RMB.

To further incentivize the agreement signing process, Yitian claimed that the unit type and public facilities of Mutoulong would be fully upgraded through the redevelopment (as the application of pay-off rule). Pedestrian channels connecting to the subway would be built while a 300-m-long landscape corridor and a 10,000 square metres eco-friendly public space would be constructed in the newly constructed community. Art, culture, fashion and food facilities would be designated with high-end commercial supports. In addition, the "Lang Lang Music World", an international theatre named after the prestigious piano master, would be located in Mutoulong. Yitian promised that after the redevelopment, the community would be built into a modern ecological community with the Music World as a landmark of Shenzhen.

The above publicity strategies substantially motivated property owners to be cooperative. As some cooperative residents noted, *"our apartment was old and small, so my family is excited about the redevelopment and hope to live in a new* in situ*unit as soon as possible. The compensation terms are satisfying, and the new facilities will greatly improve the current living environment when the reconstruction is completed. To support and speed up the redevelopment process, my family moved out immediately since we signed the contract."* (Interview of one resident, Dec-07–2016).

On the opposite side, some residents objected the compensation offers and they formed a coalition with other uncooperative residents to increase their influence (as the application of aggregation rule). They set up posters in key locations in the community to express their distrust to Yitian. Meanwhile, through face-to-face

[2] This refers to the floor area ratio of new to the demolished apartments.

discussion, they endeavoured to prevent the 'moderate groups' who were not resolute to relocate from signing the agreement with the developer. By doing so, they strived to minimize the effect of the developer's publicity actions (as the application of scope rule). According to the opposite residents,

> We don't trust Yitian to be the operator of Mutoulong redevelopment. Yitian doesn't have enough strength to redevelop Mutoulong since it is merely a small and local real estate company which does not rank top 100 in China. Its reputation is also inferior as some of its previous projects were accused of poor building quality. (SZBBS, 2015).

Despite the contest of the uncooperative residents, with substantial financial input and positive publicity, the developer-led model created an effective impact on the agreement collection. Around 1100 property owners had signed the agreement and over 80% of the households had moved out by August 2011.

5.3.2 Physical Arena

After the abolition of the demolition permit scheme in 2011, it had been a major concern of Yitian as it failed to obtain 90% resident support by 2011. As Yitian failed to catch up with the last train of the demolition permit, it had no option but to try securing full (100%) resident support. The impossibility of achieving complete consent in multi-actor price bargaining processes had pressed Yitian to use some physical actions as a non-price mechanism to achieve its goal. These actions structured the physical arena.

Yitian started by dropping the quality of property management and services to force out uncooperative residents. After the majority signed agreements and moved out, in August 2011, to monitor the activities of the remaining residents, Yitian arranged some staffs patrolling in the neighbourhood and installed two cameras facing towards the public square where residents often met and discussed (as the application of choice and scope rules). The staffs– threatened the active uncooperative residents to stop interfering with other residents to sign contracts (Interview of five residents, Dec-072016). Then the developer constructed barriers blocking the main entrance, destroyed water pipes, cut power lines and even vandalized all the street lamps. Simply on the basis of a handwriting letter from the head of the work station, Yitian illegally felled more than 200 trees in the community. These physical actions seriously ruined the normal living environment of the residential area but proved very little avail on forcing out the remaining residents. To resist the threat, more than 200 uncooperative residents aggregated in a restaurant to discuss countermeasures in February 2012 (as the application of aggregation rule). The head of the work station broke into and obstructed the meeting with 20 staffs (SZBBS, 2015).

At the end of February 2012, the district government approved the urban renewal unit plan of Mutoulong (as the application of pay-off rule). About 15% of the land in Mutoulong was designed to hand over to the local government for public housing and infrastructure. For the rest of the land, a new residential area would be built with

FAR as 7.22, around 4.7 folds of the original FAR. The extremely high density of the new community triggered more dissatisfaction of the uncooperative residents:

> The current plot ratio is only 1.55, but after redevelopment, it will increase about five times. There will be twenty 50-floor residential buildings, a commercial center, and a piano world. The original quiet and peaceful community will be destroyed with intensive density improvement. The traffic will be heavier, and the supporting facilities will fail to satisfy the future residents. Most importantly, a 150-m-tall residential building with 50 floors is not comfortable and unacceptable at all. (SZBBS, 2015).

To further pressurize the households who refused to sign a contract, the manager of Yitian exposed the speculative behaviours of some uncooperative residents through the public media (as the application of scope rule):

> We contact with the remaining residents almost every day via phone or house visit. Some owners refused to meet with our staff and some required unreasonable compensation such as 1:1.5 DCR or even 10 million RMB as rewards. In fact, there are three leaders in the uncooperative group. They want to share interests with Yitian. The offer they made is that they come forward to obtain agreements from the rest of uncooperative residents in exchange for 10% of the project profits.[3] Otherwise, they will obstruct the proceeding of the agreement signing. (Lei & Peng, 2012).

The developer's action negatively affected the reputation of the uncooperative residents and impaired the relationship between both parties. Yitian then deteriorated the property management service to force the uncooperative residents to move out (as the application of choice rule). In September 2012, Yitian removed the security and cleaning services in Mutoulong. As the remaining uncooperative residents noted,

> Without the garbage disposal service, smelly rubbish was piling in public places in Mutoulong. We complained about the developer's terrible actions to the local government but did not get any reply. We once spotted that the managers of Yitian banqueted several relevant government officials in a luxury restaurant. The developer must develop a corrupted relationship with the local government to evade regulatory penalty. Mutoulong has been an enclave out of the control of laws. (SZBBS, 2015).

The good relationship between the local government and Yitian protected the latter from being punished. In the absence of security service, more malignant incidents emerged in Mutoulong. In March 2013, a female property owner was assaulted by a stranger in the community. Several days later, another uncooperative resident was stabbed by two strangers (Wu & Lan, 2013). The female owners indicated:

> At 6:50 in the morning, I went downstairs and walked to my car. When I opened the rear trunk, suddenly someone started to kick me. I turned around and found that it was a strange young man with casual clothes and a travel cap. He twisted my left arm to the back and overwhelmed me when I tried to resist. I desperately called for help which alarmed our neighbours. The man rapidly rushed out of the community… The only reason for the assaulting was supposed that I have not signed the agreement with Yitian. I agreed with redevelopment but still cannot confirm all the compensation details. This ferocious behaviour is so unacceptable that I cannot trust Yitian anymore. (Wu & Lan, 2013).

---

[3] The requirement of 10% of the project profits were rejected by the uncooperative residents and the new project manager.

These vicious security incidents drew significant public attention once the public media released the news. Residents blamed the developer for discharging these terribly physical behaviours. In the face of media's inquiry, the developers said they never hired anybody to hurt the remaining residents nor did it cut off water and electricity supply (Wu & Lan, 2013). It insisted that some residents made up these rumours for higher compensation. The security incidents decayed the relationship between the developer and the uncooperative residents. The later complained that the local government was reluctant to file investigations to the security incidents and Yitian did not get punished for its cease of property management service provision. Their distrust to the local government was strengthened. To prevent further deterioration of the security situation, the district government claimed to take charge of the security service in Mutoulong.

At the meantime, the project delay triggered the early cooperative residents' dissatisfaction with their temporary rental experience. As the cooperative residents complained online,

> I'm very depressed. I had moved out of Mutoulong for more than two years. I had supposed I could move back to the new house soon. But the redevelopment could not be progressed since there are still more than 300 residents opposed to it. These remaining owners make us, more than a thousand cooperative households, cannot go back home. I really cannot figure out the reason why they don't agree with the redevelopment since the developer has already offered a very good compensation scheme. (SZBBS, 2015).

> We thought the redevelopment would be accomplished within three years, and I promised my son to get married in the new apartment. But after almost three years, the residential area is still a mess. My wife is quite old and not in a good health condition, so we have to live close to the hospital. Nowadays, landlords are reluctant to rent an apartment to us. This is because they are afraid that my wife would die in their house which may negatively affect the lease of the house in the future. (SZBBS, 2015).

> We must live nearby because our child is studying at the elementary school next to Mutoulong. Renting house outside is very inconvenient. Over the past two years, we have been forced to move three times. Since the house price is rocketing in Shenzhen from 2009 to 2012, despite the rental subsidy from Yitian, it becomes mountingly difficult for us to rent an adjacent appropriate unit. Now our whole family is looking forward to moving into the new house as soon as possible. (SZBBS, 2015).

To further exert pressure on the remaining uncooperative residents, Yitian and the contracted owners formed a coalition (as the application of aggregation rule). Yitian requested the cooperative owners to sign an engagement letter, in which they agreed to empty their units for Yitian to dismantle the windows and doors. On completion of that, with the support of the cooperative residents, Yitian recruited a third-party assessment company to evaluate the buildings as decrepit housing and then delegated a demolition company to demolish the buildings. According to the cooperative residents:

> Uncooperative residents are greedy and selfish. The only thing they want is higher compensation generated from the project. The minority should obey the majority since most of the affected residents have agreed on the redevelopment. Their behaviors hindered the redevelopment and did damage to the public interest because the redevelopment is a collective business. Though the developer conducted some violent behaviors, it was just willing to speed up the redevelopment. (Interview of three residents, Dec-07–2016).

The collectivism norm added moral superiority to the cooperative residents and their strategic action. In December 2012, some retired residents organized a property owner association and engaged on site to persuade the uncooperative residents to stop their resistance (as the application of aggregation and choice rules). However, the communication was ineffective according to the head of the property owner association,

> The property owner association, as a representative, went to discuss with the opposing residents. In the communicating process, some of them asked for 10 million RMB compensation, some were very impatient and reluctant to communicate and some even took a knife to intimidate us. In the online forum, they stigmatized us as gangsterdom and slandered us breaking into their house and harassing their normal life. (Interview of members of property owner association, Dec-07–2016).

As the tension among the local residents was accumulated, on March 26, 2013, more than 60 contracted residents returned to the site to pressurize the uncooperative residents. The communication triggered violent conflict and some even physically assaulted each other (Jing, 2013).

For the opposing residents, they formed a "property owner liaison group" to mobilize the remaining residents against the developer (as the application of aggregation rule). They invited some TV programs and local newspapers to publicize their resistance to the project, and complained the developer's behaviours as "cruel" and "fierce". They also seized the opportunity of social media and posted online diary to denounce the violent and vicious behaviours of the developer. The deployment of information rules had generated public sympathy which devastated the developer's reputation (Huang, 2013).

Through the online forum, remaining residents expressed their oppositions in following perspectives:

Some residents resisted the redevelopment out of their distrust to the developer,

> We are terrified by Yitian's vicious behaviors and won't trust such a company any longer. As it is not a trustworthy developer, even if the DCR 1:3 is meaningless. What should we do if Yitian has placed the relocated buildings in the worst location? What should we do if the newly built housing quality is not guaranteed? Without answering these questions, we will not sign contracts. (SZBBS, 2015).

Some residents were against the coalition between the cooperative residents and the developer:

> The developer and the cooperative residents formed a coalition to intimidate us by an array of illegal behaviours. I respect the choices of the cooperative residents and agree that if they intend to demolish their buildings which are recognized as decrepit buildings. But why don't they respect my choice? Why should we sign contracts as long as they have signed? We have our rights not to sign contracts because we have property rights for our properties. (SZBBS, 2015).

Some residents firmly believed there was a corrupted relationship between the local government and the developer, and they were expecting Yitian and the corrupted officials to be punished.

> The mistakes made by Yitian was partly due to the connivance of the local government. The government did not prevent the developers from taking vicious behaviours until personal injury happened. Why was the official who allowed the tree felling not punished? Why is there no one responsible for the cleaning and public security in Mutoulong? If Yitian has not been punished for its illegal activities in the past several years, the dark curtain of Mutoulong redevelopment will not be lifted. Mutoulong redevelopment will be an example for the future residential redevelopment in Shenzhen. If the following developer performs as what Yitian has done, the authority of law in Shenzhen will be completely absent. (SZBBS, 2015).

Some residents suspected the legality of Yitian in Mutoulong redevelopment based on their misunderstanding of the relevant policies, *"after the introduction of the national policy in 2011, the demolition operator cannot aim for profit. The redevelopment case in Mutoulong definitely contradicts to the regulation so all actions from Yitian are illegal. Yitian is only a candidate not the final developer. How can it demolish buildings?"* (SZBBS, 2015) However, Yitian was not equivalent to a role of the demolition operator in Mutoulong redevelopment.

In the face of the escalating conflicts in Mutoulong, the local government sanctioned the tree felling behaviour of Yitian but still felt reluctant to get involved. As the official explained, *"some conflicts were just contradictions between the neighbors. It is not appropriated for the government to intervene. Yitian should be punished for logging trees without legal procedures. But the demolition of the houses was private agreement between the property owners and the developer, which did not break the law. For the villainous assaulting incidents, there is not any evidence to prove Yitian should assume the responsibility. Now, both parties of the owners come to the government to petition, we cannot intervene due to the constraint of the upper law. What we can do is to require Yitian to provide a guaranty. This will help to remit residents' concern about redevelopment accountability."* (Interview of LXY, Nov-28–2016).

The physical actions of the developer did not resolve its conflict with the uncooperative residents but brought the bargaining process into a halt. From September 2011 to November 2013, only fifty-five households agreed to sign in addition to existing consents. In November 2013, under external and internal pressure, the manager of the Mutoulong redevelopment project resigned.

5.3.3 Negotiation Arena

To reduce the tension between Yitian and the uncooperative residents, in 2014, Yitan hired a new manager to lead the project. By establishing an effective negotiation arena, the negotiation practice between Yitian and uncooperative residents resumed, which improved the communicative efficacy. Quoting the new manager:

> I'm experienced in implementing redevelopment project since I used to work in the redevelopment office of the Longgang District Government. The remaining residents have great antipathy towards the developer because Yitian did some rude and vicious behaviours in the past. The former manager did not have any experience in redevelopment projects. His

> attitude was arrogant and the way he handled the project was not appropriate. I personally don't understand the reason why he escalated the confrontation between the developer and the residents. Conflict is not a solution to the project delay and the accomplishment of the project relies on the coordination of the developer, the property owners and the local government. (Interview of the new manager of Mutoulong redevelopment, Jan-08–2017).

Once the new manager took charge, he strived to improve its employees' communication skills. He required the staffs to transform from a condescending attitude to an equal position and make efforts to create a win–win situation in the negotiation process. As many of the remaining property owners were grieved over Yitian, the staffs sincerely communicated with them to gain their trust and eliminate their grievances (as the application of information rule). As the manager indicated,

> The most critical task is to do the basic work well, which is to keep a good relationship with the residents. By in-depth investigating their situations and requirements, we could comprehend the common as well as personal requirements. It is inappropriate if we only solve common problems. For some residents suffering from poverty, we could properly compensate more subsidy. The households living in the newly renovated houses can obtain more compensation for interior decoration than living in the unfurnished houses. The redevelopment will be continued only when everyone is satisfied. However, claims beyond the appropriate standard will never be met for the remaining residents. The DCR is the bottom line. (Interview of the new manager of Mutoulong redevelopment, Jan-08–2017).

Since the strategies changed, the relationship between the developer and the remaining uncooperative residents has been improved, which accelerated the agreement signing process. Yitian sent gifts to all residents at every major festival and shared redevelopment progress through social media and texts. It no longer incited contradictions between the cooperative owners and the uncooperative owners and sometimes even resolved the conflict between the two groups of residents if any dispute occurred. It also began to provide security patrols and garbage disposal services in Mutoulong. In the interview process, the new manager denied the rumour that some residents required 10% share of project interest and supposed the rumour was made up by the former manager to damage the remaining residents' reputation.

One further finding is that some uncooperative households formed a new coalition to raise their bargaining power (as the application of aggregation rule). In 2016, two representative residents accumulated 30 "tough" households to directly negotiate with the new manager. As one of the representative Mr. Zhang indicated:

> Actually, all the remaining residents realized that the proceeding of the redevelopment relies on agreements from all the residents. So, anyone who sticks to the end will not lose out. The worst situation is that the developer quit since its capital chain breaks. If that happened, the cooperative residents might suffer greater damage since their houses had been demolished. For the uncooperative residents, they still have houses to live. In recent years, as the housing price has climbed for several times in Shenzhen, the developer is still able to seek huge profit in this project. As long as the developer has a profit margin, it won't easily give up and has to negotiate with us, the remaining uncooperative residents. (Li & Feng, 2016).

As he has lived in Mutoulong for several decades, Mr. Zhang developed a harmonic relationship with his neighbours. Under his coordination, a group was formed with the support of other 28 households. Mr. Zhang and another representative together negotiated with the developer and mutually supervised each other. The greatest contradiction was that residents were required to move out immediately after they signed the contracts, but the residents were willing to relocate until the 100% of agreements was accomplished. After more than eighty rounds of communication, Yitian compromised and agreed to the households' request. Said one of the representatives, *"this cannot be achieved if we bargain with the developer individually. The coalition we formed strengthened our powerand drawdown enough patience and sincerity from the developer. We felt enough respect in the negotiation process, and that's the key to the agreement signing"* (Interview of residents, Dec-07–2017).

Once these thirty households signed the agreement with the developer, their interests were bundled with the developer and their attitudes subsequently changed. Given that they had the similar feeling and experience with the remaining residents, these households who signed contracts in late period voluntarily took the responsibility to convince the remaining households (as the application of choice rule). If succeeds, they will be rewarded by the developer.

Early cooperative residents indicated that the late cooperative residents were able to obtain higher compensation, *"the double-sided contracts are very popular in Shenzhen. These late cooperative residents claimed that they had gained the same compensation with us. But that's impossible! We heard that some of them even got additional 2 million RMB for their cooperation. However, we, as cooperative owners, are very forgiving. We will not disagree with any methods as long as they can make the remaining residents sign contracts as soon as possible. If we require equal compensation from the developer, the project will never be accomplished."* (Interview of residents, Dec-07–2016).

The new manager in Yitian asserted that they deliberately restricted the double-sided contracts for the consideration of fairness, especially when it was very difficult to maintain confidentiality. *"It was normal to compensate proper amount of money to some owners who were suffering serious illness or live in poor condition. However, the DCR was a principle which cannot be changed."* (Interview of the new manager of Mutoulong redevelopment, Jan-08–2017).

After the coalition signed agreements with the developer, there were still thirty-three tough residents in Mutoulong in January 2017. The developer indicated that most of these residents are relatively poor and underprivileged. They have struggled for a lifetime in Shenzhen with the only production of this apartment, so they are willing to gain a huge return from the redevelopment. This responds to the field observation that the existing uncooperative residents wear rugged cloth and still live in the shanty area. For these left residents, their requirements and expectations are different.

Some residents refused to sign contracts out of the resent to the developer:

> Since the agenda setting of Mutoulong redevelopment, the whole project has been full of illegal issues. The most grievance of us was that the developer demolished buildings since its executor role has never been confirmed. Yitian openly broke the law but the government never punished it. As the government will seek huge benefit from the redevelopment, it won't pay attention to our interests. The residents who signed the agreements formed a group and

> came back to assault us did not receive any penalty. When Yitian hired someone to hurt my family, we called the polices but they even did not give an alert. There were a lot of monitoring cameras in Mutoulong and its surroundings. But the policy indicated it was very difficult to find the assailant. Therefore, we won't sign contracts until Yitian and the perpetrator have been punished. We cannot afford lawyers to protect our legal interests. The only thing we can do to prevent us from harm is living here and refusing to sign the agreement. We have been working for our work unit for several decades and only obtain this apartment. We can't afford to lose. (Interview of residents, Dec-07–2016).

Some residents resist to be cooperative due to a lack of trust in the local government:

> People with discerning eyes can see the corruption between the local government and the developer. Yitian was able to illegally cut down trees in Mutoulong on the basis of a hand-writing letter from the residential committee. The police station was very reluctant to take care of the security issues in Mutoulong. With only 500 m to the district government, Yitian did so many vicious things to our residents. All of these were connived with the local government. Some cadres in high position shielded the developer. (Interview of residents, Dec-07–2016).

Some residents hope the local government applies government-led approach to Mutoulong redevelopment:

> The Ludancun is able to work as a redevelopment model for the residential redevelopment. After comparing various conditions, we all know which one is superior and which one is inferior. In Ludancun, the unit type, the orientation and even the date of relocation were set before they moved out. The bidding for lower FAR is able to protect their living environment. The entire redevelopment process was under the leadership and supervision of the local government. The Ludancun mode obviously can protect our interests. As developers are all seeking profits in the redevelopment process, the local government should take the lead in drafting a relocation scheme in Mutoulong. (Interview of residents, Dec-07–2016).

Due to the different requirement, these remaining residents didn't construct a formal coalition to negotiate with the developer but the communications between them were frequent. When conducting the second site visiting, some uncooperative residents indicated that they had heard about the researcher from other uncooperative residents who had been interviewed. When they heard that another 30 tough residents had signed the contracts with the developer. Some of them were outraged and abused the government and the developer.

When asked what role the local government played in the redevelopment process, the manager indicated that:

> We are expecting the government to get more engaged to persuade the remaining owners. The lack of involvement of the local government triggered residents' doubts about the redevelopment project. After all, everything in China is controlled by the government. Though the residents are always blaming the government, they still trust the government most. Now given that more than 98% of the residents have signed the agreement, we are disappointed that the government refused to interrupt. However, if Yitian ran out of financial resources and quit, the government would intervene to maintain social stability. Therefore, the market-led redevelopment is not merely a market behaviour, but also involves public interests. (Interview of the new manager of Mutoulong redevelopment, Jan-08–2017).

The officials in the Luohu District Redevelopment Bureau claimed:

> The market-led residential redevelopment was implementing for the first time in Shenzhen and even in China. As the local government, we are not familiar with our responsibility. Everything is 'to go across the river by touching the stones.' We hope residents sign agreements as soon as possible to accomplish the redevelopment, which will significantly diminish our pressure. However, we cannot get involved in Mutoulong due to the restriction of the superior laws. (Interview of LXY, Nov-08–2016).

When discussing the reason for the project delay, the manager noted that:

> The government did not do what it should have done. There is a problem with the requirement of the 100% contract, which is very unreasonable. A bottom line such as 90% or 95% must be established to eliminate the speculative behaviours of the nail households. The current policy is to encourage everyone to be a nail household, but it kidnapped the interests of the majority residents. Under no circumstance will any residential redevelopment be completed in accordance with such a requirement. The policies position the developer as a stronger player and the residents as vulnerable groups so that they tend to protect the rights and interests of the owners. From the perspective of financial resources, material resources, and even social influence, individual owners are indeed very weak compared with developers. However, the provisions in Shenzhen have infinitely amplified the power of the owners and indiscriminately encouraged them to maximize their personal interests by opportunistic behaviours. (Interview of the new manager of Mutoulong redevelopment, Jan-08–2017).

Since switching strategies, more than 150 "tough" residents had signed the consent. By September 2017, there were still more than 20 households remaining opposing the redevelopment associated relocation. The most critical requirement of these residents, according to the manager, is still higher compensation. However, the manager holds strong confidence about the redevelopment:

> To engage in a residential redevelopment is equivalent to engage in a battle and the developer must be well-armed. Yitian now has three magic and indispensable weapons for the success: a good team, a good strategy and harmonious relations with the local government. Our team is composed of well-trained staffs with excellent communicative skills. Through sincere communication with the residents, as a good strategy, we will relief grievance of the residents. The most important is the harmonious relations with the local government, which decides whether our project can proceed or not. Although the local government cannot get involved, we still have to respect and develop an amicable relationship with it. (Interview of the new manager of Mutoulong redevelopment, Jan-08-2017)

5.4 Exogenous Influences and Their Impacts on Power Relations

The exogenous influences that have significant impacts on power relations in Mutoulong redevelopment are composed of three types: the central policies, the local policies and attribute of the community.

The central policies had an important impact on the power relations from three perspectives: the property rights institutionalization, participatory redevelopment reform and the abolition of demolition permit institution. The institutionalization of property rights legitimized and safeguarded their rights to possess and benefit from the private property rights. The legalized property right substantially contributed to the formation of structural power resources to the affected residents. After the property right reform, in Mutoulong, some households appeared to perceive their own property as a commodity with a strong conviction that they had the power to seek immense fortune from the redevelopment. Some residents held a firm belief that they were empowered to adopt resisting strategies and actions when facing unfair treatment and enforced behaviours. The participatory reform in 2009 altered the power structure in the residential redevelopment projects. It required the local government to play a regulator to restrain its power in urban redevelopment, while empowered the affected residents to take part in the decision-making process. In 2011, the abolition of the demolition permit institution formally prevented the government intervention in market-led redevelopment, which considerably weakened the power of the local government.

Local policies resulted in the power asymmetry between the developer and the affected residents. The two-round public participation promulgated by the urban renewal policy in 2009 granted substantial structural power resources to the affected residents to negotiate with the developer for fair compensation and treatment. In line with the central policy, in 2012, a new policy which regulated the developer has to obtain 100% of agreements from the affected residents in land assembly. This policy authorized the power to influence and even impede the redevelopment process to any individual household, which provided profound structural power resources to each property owner.

The influences of the attribute of the community on the local power relations have been identified, including the physical attribute, social network, values of the participants in the community. The large size and premium location of the community and the superior medical and educational facilities in the surrounding area shaped solid allocative resource for the affected residents. As most residents had been living in the community for three decades, the close social relationships between the affected residents facilitated them to form coalitions. Meanwhile, different values of the residential redevelopment resulted in the different deployment of power among the affected residents. Since the residential area was once built by the municipal government for public housing arrangement in several work units in the 1980s, despite housing reform, some residents still retained a strong attachment to the collectivism. They consequently deemed the residential redevelopment as a collective business. These residents chose to sign the agreement in the short term to support the redevelopment and even attempted to influence the choices of the remaining uncooperative residents by mobilized actions. However, the marketed-oriented reform bifurcated the sharing values in the community. As the properties have been privatized through housing reform, some residents regarded the residential redevelopment as a certain market decision which was irrelevant to public interests. For some residents, the apartments in Mutoulong were their principal asset so that they were extensively prudent about the disposal of their houses. Some residents, especially the laid-off households, treated the redevelopment as a rare opportunity to maximize compensation from the developer or the government. For their profit, they either refused to sign an agreement or adopted some opportunistic behaviours, which eventually resulted in project delay.

5.5 Endogenous Influences and Their Impacts on Power Relations

Three main stakeholders—the developer, the cooperative residents and the uncooperative residents—structured and restructured the power relations in Mutoulong redevelopment through the configuration of resources and rules. The developer's power primarily stems from both the structural and agent perspectives. It obtained development right to conduct redevelopment cases from the participatory reform and also were empowered by resources at the agent level, including expertise, capital, reputation and experiences as allocative resources and close relationship with the local government as authoritative resource. As a role of initiator, the developer commenced to structure the residential redevelopment in order to obtain agreements from the affected residents and then seek enormous profit from the project. At the onset of the project, the developer deemed itself has an equal position with the affected residents. As the financial support from national banks strengthened its allocative resources, it adopted several publicity strategies (as the application of information and pay-off rules) to collect support from the affected residents. Through frequent deployment of these rules, the developer achieved its goal to obtain documentary support from more than 70% of the households. In the second arena, considering that supports from the majority of the affected residents reinforced its authoritative resources, the developer deemed the remaining uncooperative residents as being in a lower position and changed its position from equal to dominant. Without investigating the requirement of the remaining residents, it began to adopt an array of fierce and violent actions (as the application of choice and scope rules) to compel the uncooperative households. Despite with high frequency and intensity, these rules did not work as effectively as the ones in the first arena but caused severe conflicts among the three main actors, which resulted in enormous social cost. In retreat, after realizing the strong power of the affected residents, the developer repositioned itself an equal position in the negotiation arena. It started to resort to a sincere and patient negotiation strategy (as the application of information rule) to seek collaboration with the remaining residents. So far, through highly frequent deployment of the new rules, the consenting process has been accelerated.

Residents' power is primarily originated both in the structural definition and allocative and authoritative resources. The property right and the public participation mechanism of participatory redevelopment significantly empowered the residents to negotiate with the developer and participate in the decision-making process. The premium location, large size of the land and inferior living environment of the community added up to their allocative resources. Meanwhile, residents in this case also resorted to authoritative resources (social networks, collectivism and spatial and economic awareness) and allocative resources (social and traditional media) to strengthen their power. Being satisfied with the compensation scheme, most residents signed the contract with the developer for a temporary transfer of the property right. However, the contracts between the developer and cooperative residents demised the latter party's structural resource and marginalized them in influencing the redevelopment process. Motivated by the collectivism norm, they formed coalitions with other cooperative residents and even the developer to pressurize the uncooperative residents to agree with the project (as the application of aggregation and choice rules). These rules failed to achieve their aims but intensified the internal contradictions.

Influenced by the economic and spatial awareness, the uncooperative residents held firmly the structural resource, reluctant to transfer the property rights to the developer. In the first arena, residents who were against the redevelopment formed a coalition and resorted to publicity strategies to forbid the effect of developer's publicity actions (as the application of aggregation and scope rules). However, these rules did not take effect to avoid the agreement signing from the majority. Facing violent behaviours and inferior property management services, in the second arena, these residents availed themselves of the frequent advance of public and social media (as the application of choice rule), which turned to be successful in constraining the developer's enforced and violent behaviours. Once the developer shifted its strategy in the third arena, some uncooperative residents formed coalitions based on the same value and requirement to strengthen their resources (as the application of aggregated and choice rules) to bargain with the developer. Since these rules generated effective outcomes to achieve their goals, they turned to be cooperative eventually. However, the remaining tough residents, based on their structural resource, kept refusing to sign the contract to resist the developer's strategies (as the application of scope rule), resulting in project delay.

The local government originated its power in the structural resources mainly include definition of regulations and policies, and jurisdictions on land use and development intensity. In this case, the local government primarily focuses on redevelopment project approval to confirm the developer and redevelopment plan setting which influenced the distribution of interests between the developer and the residents (as the application of position and pay-off rules).

5.6 Outcome and Evaluation

5.6.1 *Outcome*

Negotiations between the developer and the uncooperative residents continued at the time of writing (2018). In this DRAR project, the proposed floor area ratio (FAR) is 7.22, which is far higher than the existing ratio of 1.55. Since the majority of the residents signed the consent and moved out, Yitian has been paying their rental cost. The cost has accumulated to 600 million RMB up to the end of 2017. The cooperative residents have suffered by living in temporary accommodation, while the uncooperative residents have to cope with poor living conditions. Both sides of the affected residents had complained to the government hoping for solutions, but both were disappointed because a lack of policy alternative given the new market approach remains effective. The local government's reputation was harmed. The new market approach in Shenzhen has led to a "triple-lose" outcome causing severe social cost and loss.

5.6.2 *Equity Evaluation of the Outcome*

Yitian has invested heavily but is expecting an enormous economic gain by new apartment sales. This expectation seems not unrealistic given that apartment prices in Shenzhen has been skyrocketing. For the property owners, without any investment, their living conditions and property value are expected to improve greatly. The government is also expecting high gains in this process as 15% of the land will be transferred free of charge from Yitian for the construction of public facilities. Besides, the urban environment will be significantly improved at a low relative cost to the government.

Despite the optimal benefit scenario, the distribution of project returns is probably not in line with the equity principle that those who receive greater benefits are expected to pay more (Imperial, 1999). In the absence of detailed calculation taking into account the unpredictable hiking rates of land values, infrastructure costs and the expenses on administrative tasks, there is no concrete numbers but speculations about the winners.[4] On top of the doubt is whether the developer and the local government invested in proportion to their expected gains. Dramatic increases of market housing price provide evidence to support this speculation about the developer's enormous gain, while the money value of 15% of free project land is probably difficult to be matched by government contributions. More concrete data from fieldwork in Mutoulong indicates that residents are not treated equally. Yitian had to sign a "double-side contract" with some "tough" households to satisfy their request for higher compensation. A double-side contract comprises a formal contract on one side, identical to excising consents and a hidden contract on the other, with a higher pay-out. This apparently violates the equity principle for resource allocation and distribution.

[4] In theory, there should be detailed analysis and calculations on the financial flow, as government approval for the redevelopment proposal and bank approval for loan applications would all need such details. However, none of the finance details are transparent. Instead, each stakeholder has its own assumptions and calculations, leading to the various speculations and different projections in the negotiation of compensations.

5.6.3 Institutional Adaptation of the New Power Structure

In Mutoulong, all residents exercised their power of choice in dealing with contracts. After realized that on-site posters had limited impacts on affecting other property owners' decision of agreement signing, they resorted to social media and public media and some mobilized actions to pressurize the developer in the next arena. However, after the abolition of demolition permit institution, each household has certain power to influence the project. The remaining uncooperative residents seldomly adopted collective actions but kept refusing to sign agreement for achieving their objectives.

For the developer, as the agreement signing process was smooth and fast in the publicity arena, it failed to realize the empowerment of the residents and the change of power structure. In the physical arena, the developer took dramatic actions to influence residents' choice (between stay and relocation). This resulted in further tension between the developer and unsigned households and severely delayed the agreement signing process. The developer drawn lessons from the physical arena and changed its strategy in the negotiation arena and thereby facilitated the agreement signing process. Realizing that excessive expansion of property owner's power was the main cause for project delay as bargaining cost, the government had amended the market approach in 2014 in its urban redevelopment policy.[5]

5.6.4 Transaction Cost and Project Delay

Transaction cost is incurred in the Mutoulong project for obtaining residents' approval of the redevelopment proposal. The cost is tied with project delays because manpower cost and bank interests paid for the loans all increase in time. Field evidences show that transaction costs in Mutoulong are associated with power relations and stakeholders' behaviours.

Analyses of the power relations in the three arenas show that the contract signing process went through different power symmetry environments and resulted in varied outcomes. In the publicity arena, thus, one developer with plentiful of allocative and authoritative resources balanced out many residents who had limited resources but one single resource (i.e. the consent to sign) with infinite weightage. Both power-holders had equal positions in negotiation thus the agreement signing process was fast and smooth. In the physical arena, more power skewed to the developer side as residents who signed off the agreement changed side in the power balance. Nevertheless, the effect of this shift of side on power symmetry was negligible because residents who did not sign the agreement held infinite power. The power symmetry was eventually broken by a change of the developer's position from equal to dominant and

[5] Refer to footnote 1. The amendment does not affect Mutoulong as it has followed the market approach. In order to progress the redevelopment, Yitan needed to achieve full agreement by all affected residents.

the associated mobilization of further resources to compel the uncooperative households. This led to power asymmetry which resulted in more resistance thus slower agreement signing process. In a retreat, the developer repositioned itself by refraining from the use of aggressive resources in the negotiation arena, thus reinstalled power symmetry to the process. The adjustment strategy has so far been working well to accelerate the agreement signing process.

Opportunistic behaviour can be observed for both developer and the residents in Mutoulong. To encourage residents to surrender their property rights, Yitian arranged loans for providing residents with allowances to pay for their temporary housing until the project was completed. The substantial investment changed the asset specificity by adding to the financial burden, thus generated new pressure for the developer to replace the "carrot" approach (i.e. give incentives for those who signed the agreement) to a "stick" approach (i.e. to make the living environment harsh by stopping services). These opportunistic actions destroyed the residents' trust and caused project delay. On the other hand, some opposing residents knew that the developer was keen to complete the project in the shortest timeframe, which could add to their bargaining power. So, they refused to sign off contracts unless their demand for a higher level of compensation was met. The opportunistic behaviour of the residents to repeatedly bargain further delayed the project progress and increased the transaction cost. High transaction costs resulted in low efficacy of the redevelopment progress and sub-optimal distributional outcome.

5.7 Discussion

This section discusses the advantages and disadvantages of developer-led power relations in delivering an urban redevelopment project. Market-oriented reforms have removed supply-side constraints for developers and allowed residents to participate in decision-making, thereby enabling them to capture a greater share of redevelopment benefits. However, in Mutoulong, Shenzhen's new market-driven approach—characterized by policy gaps and high transaction costs—hindered rather than facilitated the redevelopment process. The lack of unanimous support from residents was directly responsible for significant project delays, while the developer's approach to project planning and implementation, coupled with the absence of government facilitation—particularly when tensions between the developer and residents escalated—exacerbated social losses.

Three factors further explain this failure. First, the new market-oriented approach did not adequately include residents in project planning. The redevelopment plan and the terms of the relocation agreement were drafted solely by the developer, without meaningful engagement with the households, who were merely given the option to accept or reject the agreement. This issue reflects a broader pattern in participatory practices in China, where limited political will to engage the public combines with a widespread belief that the government and technical experts are better positioned to make decisions (Enserink and Koppenjan, 2007; Tang, Wong and Lau, 2008). Second, the policy lacked an effective mechanism to control project delays. Since the cost of delay is substantially higher for the developer than for the uncooperative households, the latter were able to use project delays as a bargaining strategy to secure higher compensation. Neither the developer nor the local government appeared to have a clear exit strategy to manage such delays. These findings agree with Schoon (2014a) that the loss of regulatory power in China's urban redevelopment increased informal activities and increased temporary systemic instability. Third, administrative conflicts between different tiers of government also contributed to project delays. Following the central government's abolition of the demolition permit, developers are required to secure the consent of 100% of residents. The local government no longer possesses the authority for compulsory land acquisition in housing redevelopment. This central-level policy change was not effectively aligned with related regulations, substantially increasing transaction costs. The absence of a clear institutional framework can impose extremely high costs in urban redevelopment processes.

5.8 Summary

This chapter examines the power relations in the developer-led participatory redevelopment process, as summarized in Table 5.1. The analysis indicates that the developer assumes an initiating role, negotiating with affected residents to achieve land assembly. Residents are divided into two main groups: those who support the developer and those who remain uncooperative. Policy inconsistencies between the central and local levels empower individual households to halt the redevelopment simply by refusing to sign agreements with the developer. This arrangement grants residents considerable influence over the decision-making process. Cooperative residents who have signed agreements transfer their property rights to the developer and are consequently marginalized, while uncooperative residents exert substantial power to strategically negotiate higher compensation, thereby prolonging the negotiation process and delaying the project. The evolution of power relations in this participatory redevelopment follows three stages. Initially, interactions are collaborative and inclusive, but they quickly deteriorate into conflict and disorder before ultimately reverting to a cooperative arrangement. Although the final collaborative phase facilitates negotiations, persistent power asymmetries continue to pose significant obstacles to the successful completion of the project.

Table 5.1 Summary of power relations in the developer-led case

Actor	Rules employed	Resources mobilized
Phase 1: Publicity Arena (2009–2011)		
District Government	• Position rule • Pay-off rule	• Structural resources
Yitian Group (Developer)	• Pay-off rule • Information rule	• Allocative resources • Authoritative resources • Structural resources
Cooperative Residents	• Choice rule	• Structural resources • Authoritative resources
Uncooperative Residents	• Aggregation rule • Scope rule	• Structural resources • Authoritative resources
Equity: Mixed Adaptability: High Transaction Costs: Moderate		
Phase 2: Physical Arena (2011–2013)		
Yitian Group	• Choice rule • Scope rule • Position rule • Aggregation rule • Information rule	• Allocative resources • Structural resources • Authoritative resources
District Government	• Pay-off rule	• Structural resources
Uncooperative Residents	• Choice rule • Aggregation rule	• Authoritative resources • Structural resources
Cooperative Residents	• Aggregation rule • Choice rule	• Structural resources • Authoritative resources
Equity: Low Adaptability: Low Transaction Costs: High		
Phase 3: Negotiation Arena (2014–2017)		
Yitian Group (New Manager)	• Information rule • Pay-off rule • Choice rule	• Allocative resources • Authoritative resources • Structural resources
Late-Cooperative Residents	• Aggregation rule • Choice rule	• Structural resources • Authoritative resources
Remaining Holdouts	• Scope rule • Position rule	• Structural resources • Authoritative resources
Equity: Improved but inconsistent Adaptability: High Transaction Costs: Reduced		

References

Huang, X. (2013, April, 16). The affected residents were hurt because of redevelopment in Mutoulong. *Shenzhen News*. http://www.sznews.com/news/content/2013-04/16/content_7940259.htm

Jing, B. (2013, Apr–10). The cooperative and uncooperative fighted with each other in Mutoulong. *News.ifeng.com*. http://news.ifeng.com/gundong/detail_2013_04/10/24048425_0.shtml

Lei, D., & Peng, S. (2012, May–16). The refusal to be relocated by the 10% of uncooperative residents hurted the 90% of cooperative residents.*Shenzhen Evening Newspaper*. https://finance.ifeng.com/roll/20120516/6465322.shtml

Li, Y., & Feng, S. (2016, Dec–12). Why did the tough uncooperative residents sign agreement with the developer in Mutoulong? *Southern Metropolis Daily*. http://www.oeeee.com/html/201612/12/437508.html

Lu, Y. (2010). The demolition compensation ratio in Mutoulong is 1:1.3. More Than 1000 Households Signed the Agreement. *Nanfang Metropolis Daily*.β http://sz.house.163.com/10/1228/09/6OVV88PH00073T3P.html

Schoon, S. (2014). Chinese strategies of experimental governance. The underlying forces influencing urban restructuring in the Pearl River Delta. *Cities, 41*, 194–199. https://doi.org/10.1016/j.cities.2014.01.008

SZBBS. (2015). The diary of uncooperative residents. *SZBBS*. https://szbbs.sznews.com/thread-1155296-1-1.html

Wu, J., & Lan, J. (2013, Apr–16). Strangers with black coat stabbed the uncooperative residents in Mutoulong. *SZNEWS*. https://szbbs.sznews.com/thread-1803723-1-1.html

Chapter 6
"Owner-Led" Participatory Redevelopment in Nanhuacun

6.1 Introduction

This chapter investigates the power relations among the main stakeholders—the affected residents, the local government, the developer and real estate agents—in an owner-led redevelopment project. Section 6.2 provides an overview of the case study project. Section 6.3 examines the social practices and evolving power relations throughout the redevelopment process. Section 6.4 analyses the impacts of exogenous influences, while Sect. 6.5 considers the effects of endogenous influences. Section 6.6 presents the spatial and social outcomes of the project and evaluates the associated power relations. Section 6.7 concludes the chapter.

6.2 Case Context and Community Attribute

Nanhuacun is a large dilapidated residential area in Futian District, as shown in Fig. 6.1. The site covered approximately 161,900 square metres and comprised 70 seven-storey residential buildings, accommodating a total of 2566 households. The original FAR of Nanhuacun is 1.45. As one of the largest welfare housing projects in Shenzhen, Nanhuacun was developed by the local government in 1985 and subsequently allocated to employees in the public education, commerce, public security and engineering sectors. Following China's housing commercialization reform,

---

The term "owner-led" is used in this chapter in order to be consistent with discussions presented in the literature review and research method chapters. Through literature review in Sect. 1.2, the new pattern of owner-led redevelopment was identified. It then guided the case selection and research design in Sect. 1.3. However, on completion of the fieldwork and empirical analysis, it was clear that the owner-led redevelopment did not happen since homeowners in this case had a very passive power in deciding whether or not to join the redevelopment, rather than to implement the redevelopment. Therefore, I marked it as "owner-led" with quotation marks.

X. Li, *Participatory Redevelopment as Contested Arenas in China*,
https://doi.org/10.1007/978-981-95-9394-1_6

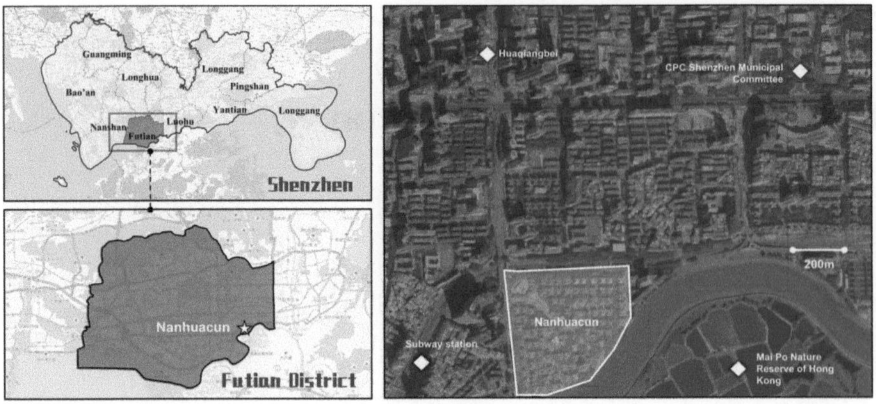

Fig. 6.1 The location of Nanhuacun, Shenzhen

the households purchased full property rights from their asset-holding work units, rendering the apartments market transferrable.

Nanhuacun occupies a highly advantageous location, in close proximity to Huaqiangbei, Shenzhen's major commercial and electronics hub. The community is well served by public facilities, with two metro stations, two middle schools and a reputable hospital all located within a 500-m radius. In addition, Nanhuacun enjoys an exceptional landscape setting, directly facing the Mai Po Nature Reserve in Hong Kong across the Shenzhen River, which further enhances the site's commercial appeal. In 2010, the Guangdong provincial government and the Hong Kong government signed the Framework Agreement on Guangdong–Hong Kong Cooperation, which proposed the development of an education and technology centre in the Lok Ma Chau Loop area. Located approximately 1 km from this planned cross-border development, Nanhuacun's site value has been further reinforced by these strategic initiatives.

Although Nanhuacun was built only 30 years ago, the physical condition of its buildings has deteriorated severely. Since 1996, residents have repeatedly petitioned the local government regarding poor construction quality and associated safety hazards. Following several on-site inspections in the late 1990s, the Shenzhen Decrepit House Appraisal Committee identified six residential buildings as structurally deficient and in need of strengthening or reinforcement. By 2000, these six buildings were officially classified as dangerous housing. Over the subsequent three years, three buildings were structurally reinforced, while one was completely demolished and redeveloped as an 18-storey high-rise residential building. The remaining two buildings, however, were left unimproved due to persistent disagreements among property owners.

After more than three decades of continuous use, the overall physical condition of the community has become severely dilapidated, with long-standing and significant safety concerns. As one long-term resident who has lived in Nanhuacun for 30 years explained:

> A large amount of sea sand and yellow clay were used during the construction process in the 1980s, which led to corrosion of steel bars and leakage of walls. Due to the lack of effective construction supervision, both the steels and the floor slabs were of deficient quality. After residents moved in, some households removed load-bearing walls to extend their usable areas, which deteriorated the structural integrity and stability of the buildings. According to relevant experts, such a building cannot withstand even a minor earthquake. We have worried about it for decades. (Interview of one resident, Jan-03-2017)

In addition to poor building quality, outdated community facilities have imposed substantial constraints on residents' daily lives. The absence of elevators in the seven-storey residential buildings has significantly affected elderly residents, while the limited availability of parking spaces has become increasingly incompatible with rising levels of private automobiles onwership. Moreover, inadequate fire-safety facilities and insufficient emergency exits pose serious safety risks to the community.

As a large-scale residential area comprising more than 2,500 households, Nanhuacun is jointly managed by two property management companies: Wanxia Property Management Company and Urban Construction Management Company. As households with relatively stronger economic capacity have gradually moved out, an increasing proportion of current residents consists of tenants employed in Huaqiangbei and surrounding areas (interview with Mr. Wu, January 5, 2017).

6.3 Power Relations Analysis in Nanhuacun Redevelopment Project

6.3.1 Owner-Led Arena

Just 15 years after its construction, by the year 2000, Building 7 had already suffered severe deterioration in quality. The walls were cracking with the rusty steels exposed, which caused serious safety concerns for the 28 households. These property owners thus negotiated with the local government for the redevelopment of their building, forming the owner-led arena.

Out of the dissatisfaction of the building quality, Mr. Wu, an experienced engineer who lived in building 7, noted,

> The floor slabs were too thin, and the concretes were not robust enough because of the mixture of sea sand in the construction. The rust would be falling from the ceiling in the downstairs when residents lived in upstairs were stomping. When a strong wind came, the whole building was shaking. We cannot tolerate this dangerous building anymore and are concerned about the collapse of the building every day. However, most of my neighborhoods are communist party members who are significantly impacted by Chairman Mao. They were grateful for the accommodation the government provided when they came to Shenzhen and dared not be selfish to require too much from the state. Different from them, I'm a professional engineer who took charge of the construction of the tallest building in Shenzhen. According to my observation, I can make a judgement that our residential building in Nanhuacun was incapable of resisting even a minor earthquake. It is too dangerous. (Interview of Mr. Wu, Jan-5-2017)

He hence attempted to guide the affected households to lodge an appeal to the local government for building redevelopment (as the application of aggregation and choice rules). Based on his reputation as an authoritative resource, Mr. Wu organized several collective meetings with the affected households in building 7 to exchange information and discuss the appropriate strategy. During the meetings, Mr. Wu repeatedly required the affected residents to take a cooperative attitude with the local government (as the application of choice rules). As he presented,

> I've experienced a lot of events led by the government such as three-anti and five-anti campaigns and cultural revolution since the foundation of the People Republic of China. So, I comprehend the Communist Party very much. If we can convince the government with solid evidence, we may achieve our ends. But we would get nothing at any time if we were against the government. I tell my neighborhood to follow my discipline and don't escalate the conflict with the government. They can politely complain to the government by calls and letters but cannot put an illegal parade or even physical fight with the officials into practice. (Interview of Mr. Wu, Jan-5-2017)

After the meetings, these residents agreed to Mr. Wu's requirement and promoted him as a representative to publicly express their opinions to the municipal government. On completion of the petition letter, he allocated it to all households for their signatures and submitted it to the municipality. As the similar issue contemporarily emerged in several residential areas built in the same period (such as Ludancun), the Shenzhen municipality attached great importance to the issue and promised to conduct a test on the quality of the residential buildings. Several months later, six residential buildings in Nanhuacun were identified as dangerous housing.

However, due to the limited financial resources at that time, the municipal government was reluctant to conduct the demolition and reconstruction but proposed a plan to reinforce the supporting structure of these buildings. Once the decision was made, the municipal government held a meeting in Nanhuacun to persuade the affected residents. This decision aroused acute opposition and conflict. Some emotional residents crushed into the meeting and pressurized the officials with fierce words. As the representative, Mr. Wu, on one hand, tranquilized the residents that the municipal government had realized the residents' concerns and planned to take actions to address the issues. On the other hand, he required the government officials to organize a hearing with the professionals who made the decision. Mr. Wu planned to persuade these professionals for redevelopment.

To relieve the fierce grievance, several cadres of the municipal government and the construction bureau came to Nanhuacun for the hearing (as the application of information rule). However, some irritated residents started to argue with the officials once the hearing commenced, resulting in chaos and conflicts on site. Mr. Wu dismissed the displeased residents and negotiated with the cadres personally:

> I once worked with some of these cadres, so they know my characteristic and personality. I explained that we never intend to create instability for the government but feel unenthusiastic to accept the reinforcement plan. These officials are working for the public. If we treated them respectfully and convinced them with substantial evidence, they would accept our opinions. (Interview of Mr. Wu, Jan-05-2017)

The proposal of Mr. Wu was mainly about three perspectives. At first, the accountability of the current structure was the main concern. The main structure of the building was very weak since it was constructed with sea sand and incompetent steels. The reinforcement of the building structure would add more columns and thus lay severe pressure on the current structure, which might deteriorate the building quality. Secondly, the proposed columns would occupy the usable area and decrease the internal space for the residents. As the relocation was necessary for the implementation of the reinforcement program, the residents preferred to demolish the building and construct a new building with better quality. Thirdly, if the reinforcement solution was chosen, patches would be all over the building to fix the cracked wall. These patches would destroy the artistic image of Nanhuacun and further degrade the reputation of Shenzhen, as Nanhuacun was built by the Municipal government.

After the negotiation, the municipal government promised to undertake an internal discussion for the redevelopment plan but had not responded to the residents for several months. As the residents in building 7 noted, *"if we residents were not pushing, the local governmentwould hold a very passive attitude. For them, the less trouble, the better."* (Interview of residents in Nanhuacun, Jan-05-2017)

With repetitive petitions from the residents, the municipal government eventually determined the redevelopment approach and delegated the government-owned development companies Urban Construction Development Company (UCDC hereafter) to take in charge (as the application of position rule).

However, without any negotiation about the redevelopment details, since its involvement, the UCDC began to encourage the residents to move out (as the application of choice rule). According to Mr. Wu, *"if the household moved out within one month, they would be awarded 4000 RMB. But this was very suspicious. If the developer was willing to demolish and reconstruct the building, why not negotiate the compensation scheme with us?"* (Interview of Mr. Wu, Jan-05-2017)

Because of developer's award, several households moved out. The UCDC commenced reinforcing the building without any notice. This action stirred up immense exasperation from the affected residents. Under the organization of Mr. Wu, the remaining residents resisted to relocating until the UCDC ceased the reinforcement and adopted the redevelopment approach (as the application of choice rule). As Mr. Wu noted,

> We informed the UCDC to formally embark on the negotiation about the compensation scheme with us. Otherwise, we would appeal to the municipal government which promised to demolish the dilapidated buildings. (Interview of Mr. Wu, Jan-05-2017)

Being afraid of the punishment from the municipal government, the UCDC commenced bargaining the compensation details with the affected residents. After five rounds of discussion, the developer promised to compensate an apartment with the same usable area, and to provide subsidy for the expenditure of interior decoration and rent for temporary housing until the new flats were accomplished. The appropriate compensation scheme encouraged the affected property owners to be cooperative. All the 28 households moved out in three months, and the demolition and reconstruction began operating.

However, at the end of 2002, three months before the redevelopment was accomplished, the UCDC suspended the rental subsidy and urged each household to reimburse 20,000 RMB for their newly constructed apartments (as the application of pay-off rule). This was because the new apartments were with larger usable area than the original one. Otherwise, the residents would not be allocated keys for their new home. The action triggered extreme indignation of the affected households. Some indignant residents intended to burst into the developer's company and created disturbance with violent actions. Once realized that Mr. Wu strived to soften the conflict between the residents and the developer and negotiated with the manager of UCDC on the residents' behalf (as an application of choice rule). As Mr. Wu indicated:

> Once I met the director of UCDC, both of us laughed. I used to supervise his other projects in Shenzhen. I conveyed our difficulty that we are not able to afford the 20,000 RMB reimbursement since our salary was merely around 1000 RMB per month at that time. I insisted that the usable area of the new apartment is the same as our previous one whilst the extended area was for the elevators of the high-rise building. He showed enough respect and arranged the project manager to 'take care' of me. After the negotiation, the developer satisfied our requirements and continuingly compensated the rental expenditure and the interior decoration. The payment for the additional areas was exempted. (Interview of Mr. Wu, Jan-05-2017)

Once the redevelopment was accomplished, the UCDC became the property provider in Nanhuacun. Though the affected residents eventually moved to their new apartments smoothly before the Spring Festive of 2002, the financial feasibility of the developer was impaired, which demotivated it to further redevelop other dilapidated residential buildings in Nanhuacun. On completion of the redevelopment, the developer claimed the project caused net losses and the "annoying" requirement of the affected residents was difficult to handle (Interview of Mr. Wu, Jan-05-2017). In addition, due to the absence of policies, few residential redevelopment projects were approved. Therefore, although the issues of the building quality were deteriorated in the 2000s, the residential redevelopment was seldomly conducted in Nanhuacun.

6.3.2 *Multi-Developer Competition Arena*

The enactment of the developer-led residential redevelopment policy in 2009 motivated private developers to rush into the dilapidated community for the redevelopment right. Considering Nanhuacun redevelopment an excellent opportunity to seek fortune, the existing property provider UCDC commenced collecting supports from the affected residents. In 2010, one small local real estate company named Tongtiandi and a reputable nationwide developer named Excellence Group (EG hereafter) joined the Nanhuacun redevelopment. Began thus the competition among three private developers, forming the multi-developer competition arena.

In the beginning, with only one developer in Nanhuacun, the UCDC only made telephone calls to persuade the residents to support their proposal. After the participation of the other two companies, three developers commenced competing for residents' trust and supports. They set agencies on site and delegated their staffs to send gifts house by house. Besides that, they publicized their redevelopment plan through posters, flyers and repeated telephone calls (as the application of choice rule). Based on its abundant capital as the allocative resource, the EG arranged free tours for the affected residents to visit its previously accomplished projects every weekend. According to some affected residents,

> We've received many gifts such as food, cooking oil, rice and so on from all three developers. We also visited several previous residential projects of EG. The building quality and the living environment of these projects are quite satisfying. During the tour, EG took care of our meals and provided considerate services. For Nanhuacun redevelopment, EG proposed the DCR as 1:1.2 and the rental subsidy is very appealing. It gave us an impression of being sincere and responsible, so, I signed the agreement to support EG as the operator of Nanhuacun redevelopment. It is not the final compensation contract anyway. If the case gets approved in the future, we will reconsider the offer and bargain for higher compensation. (Blog comments, SzhomeBBS, 2010)

Different from these residents who were attracted by the developers, some residents considered the publicity as harassment and thus objected to the redevelopment. At the same time, some sober-minded professional residents endeavoured to impact the residents' choices based on their authoritative resources (as an application of choice rule). Mr. Mao, a retired professor in Shenzhen University, posted several articles in the online forum:

> We should support the Nanhuacun redevelopment since the poor building quality caused great inconvenience to our life. But we need to calm down. We cannot be obsessed with the small gifts from the developers. They have to provide their detailed redevelopment plan including the potential FAR, the proposed public facilities and so on. Otherwise, even the 1:2 DCR means nothing to us. We should also be cautious of the strength of each developer. The implementation of the residential redevelopment requires the devotion of enormous time and investments. Developers without enough capital resources and experiences will not accomplish it. (Blog comments, SzhomeBBS, 2010)

To reduce the negotiation cost, all of these three developers refused to redevelop the high-rise residential building, which was redeveloped and built in the 2000s. Their plans triggered intense resistance from the affected residents. As Mr. Wu, who led the redevelopment of the high-rise building, indicated that

> All of them proposed a modern concept plan of the residential area with sufficient and convenient public facilities such as beautiful gardens, kindergartens, and parking areas, in which our building seven was excluded. This certainly resulted in immerse opposition. It would be ridiculous that Nanhuacun was filled with 35-floor residential buildings but with one 18 floor apartment building in it. In addition, if the redevelopment was implemented in the future, the demolition and construction process would be extremely annoying for the excluded residents. Therefore, it is impossible for us to support any of these private developers. (Interview of Mr. Wu, Jan-5-2017)

The signing of "the Guangdong and Hong Kong Cooperation Framework Agreement" in 2010 improved the expectation of the property owners in Nanhuacun. Some residents began to hold a wait-and-see attitude for possibly higher compensation. Because of ample financial resources and expertise, the developer EG obtained more supports from the affected residents than its counterparts. However, the fierce competition between the three engaged developers in Nanhuacun split residents' supports into three groups. In April 2010, even EG only obtained around 600 agreements from the property owners, which was far away from the target of 1800 households (Interview of the resident association, Jan-03-2017). Without any developer obtaining more than two in three supports from the residents, the Nanhuacun redevelopment approval did not get approved in 2010.

In 2011, another nationwide reputable developer Kaisa, with competent financial strength as an allocative resource, engaged in Nanhuacun. To obtain the supports from residents as soon as possible, it provided DCR as 1:1.4 and promised 80,000-yuan award to any households who supported their proposal. The participation of Kaisa restructured the competition among the involved developers. Out of the fear of the immense financial strength and strong reputation of Kaisa, three earlier engaged developers quit. However, in 2012, the municipal government prohibited the involvement of private developer from soliciting residents' supports in dilapidated residential areas. As Kaisa failed to obtain more than two in three supports from the residents, the approval of Nanhuacun redevelopment lapsed again. Considering the high sunk cost in Nanhuacun, the Kaisa still settled a temporary agency in Nanhuacun, but its actual actions had been few after the promulgation of the new regulation.

6.3.3 Agent-Led Arena

Since the prohibition of their involvement in the dilapidated residential redevelopment, the private developers had volumed down their publicity on site. The municipal government felt reluctant to initiate residential redevelopment due to the severe project delay in the approved market-led residential redevelopment cases. As the residents were seriously affected by the deteriorated residential buildings, they continuously petitioned to the municipal government through public letters, telephone complaints and face-to-face interviews for the redevelopment agenda. The contradiction was escalated between the redevelopment demand of the residents and the passive response of the municipal government. Some private agents intended to guide the residents to obtain redevelopment approval from the municipal government and seek immense profit from the process. In Nanhuacun, a real-estate-related agent organized a "resident association" to guide a so-called "owner-led" residential redevelopment. Under the motivation of the association, residents, reputable developer Vanke, the work station and the street office were involved for the negotiation of the redevelopment approval, which framed the agency guidance arena.

In September 2013, Dr. Shao, as a previous consultant of Vanke—one of the most reputable real estate companies in China, intended to seize fortune from the

Nanhuacun redevelopment. He planned to first accomplish the 100% supports from the affected residents to obtain the redevelopment approval; on completion of that, he then handed over the project to a private developer in exchange for enormous profit (Interview of Mr. Wu, Jan-5-2017). To achieve the aim, he banqueted and earned trust from some prestigious residents in Nanhuacun, such as Professor Mao and Mr. Wu. The trust of influential owners significantly benefited his reputation. In the next month, Dr. Shao hired several active residents to form a "resident association[1]" and pronounced it as a voluntarily organized organization by the affected residents. He bought a first-floor apartment in Nanhuacun and renovated it into an office for the organization.

As it hired some prestigious owners as consultants, in November 2013, the resident association, under the name of Vanke, claimed to guide the affected property owners for the application of redevelopment approval in Nanhuacun. The resident association held a public meeting and promulgated their extraordinary generous compensation schemes to the affected residents (as the application of pay-off, information and choice rules). The DCR would be 1:1.4 with larger usable area of the newly constructed apartment than the original one.[2] On the completion of the redevelopment agenda, each household would be rewarded 100 thousand RMB for their relocation. The rental subsidy and compensation for the interior decoration were also higher than the normal standard in Shenzhen while the relocated buildings were also promised with premium location and apartment type.

In December, to legally invite Vanke to participate in the Nanhuacun redevelopment, the resident association prepared a questionnaire of redevelopment support for the affected residents. They enlisted 20 active residents to collect supports and signatures in the community. The association also established online forums and social media groups to share information with the households (as the application of information rule). Meanwhile, due to his social influence as the authoritative resource, Dr. Shao utilized the local public media to publicize the "owner-led" redevelopment approach in Nanhuacun,

> The owner-led approach is the most ideal for Nanhuacun redevelopment since this kind of redevelopment was implemented in the community more than ten years ago.[3] Nowadays, the residents voluntarily organize an organization – the resident association – to take the local government's responsibilities for supports collections and developer invitation. The association will work with the developer to establish detailed compensation and relocation schemes and apply the redevelopment approval to the local government. This is supposed to be the most efficient and appropriate way for the dilapidated residential redevelopment in Shenzhen. (SzhomeBBS, 2013)

Through telephone and house-to-house visit, within two months, more than 95% of the property owners claimed to support the Nanhuacun redevelopment and brought

[1] Resident association (Ye Zhu Zhi Jia) is a commercial group organized by Dr. Shao. It worked as an agency to guide the affected residents to apply Nanhuacun redevelopment to the local government.

[2] To obey the regulation that the maximization DCR was not able to be higher than 1:1.3, the owner association declared to sell the addition area to the original residents in an extraordinary low price to achieve the 1:1.4 DCR.

[3] It refers to the owner-led redevelopment led by Mr. Wu.

comfort to Vanke's involvement. However, the supports which the resident association collected were different from the documentary supports required by the market-led redevelopment. The former only includes the name of the households and a contact number while the latter refers to a copy of property certificates and a letter of authorization. In March 2014, without documents as the formal support, the resident association claimed to the residents that the substantive redevelopment of Nanhuacun had started and the president of Vanke had agreed to assist in the agenda setting of Nanhuacun. In the next month, to earn residents' trust, the resident association organized a tour to the Headquarter of Vanke in Shenzhen. After these publicity actions, more than 98.3% of residents signed the invitation letter prepared by the resident association to support the invitation of Vanke to implement Nanhuacun redevelopment (as the application of choice rule).

Meanwhile, the resident association complained about the requirement of 100% agreement, *"although 98.3% of the owners agreed to the Nanhuacunredevelopment, it is almost impossible to reach the 100% requirement. The two households on the third floor of building 13 have been vacant for more than 20 years. We have never seen the owners or any tenants. Who should we refer to for the agreement?"* (Interview of the resident association, Jan-03-2017)

Since the abolishment of the market-led approach in 2012, the responsibility of support solicitation was arranged to the street office, the grassroot governmental organization. On completion of the 98.3% agreements, the resident association applied to the street office for redevelopment agenda but got rejected. As the director of the street office noted, *"for any association like the residential association in Nanhuacun, the solicitation of supports will present few difficulties if they offer very attractive compensation terms. However, the subsequently contract signing process would be very complicated since it is related to the individual requirement of compensation. Therefore, the local governmenthas to be very cautious about redevelopment agenda setting. As a governmental institute, we don't recognize 98% of agreement in Nanhucun. First, the resident association is not the property owners committee, the legally recognized organization of the property owners. It cannot assume the role of the residents' representative. Second, according to the related policies, the dilapidated residential redevelopment should be led by the local government. The current resident association is not a government organization so that the collected support will not be recognized by the government. Third, the supports collected by the association are different from the formal supports in the developer-led redevelopment. There is not any contract between both these two parties. So, it is very difficult to approve the Nanhuacun redevelopment agenda on the basis of the supports accumulated by the association."* (Interview of an official in Nanyuan Street office, Jan-12-2017)

To further pressurize the local government for redevelopment approval, in March 2014, the resident association invested more than 1 million RMB (as the allocative resource) to initiate a structural safety and reliability test of the residential buildings in Nanhuacun (as the application of choice rule). The test was conducted by the same institute (the Shenzhen Institute of Building Science) and the same standard[4] with Ludancun. The result showed that 32 buildings were identified as decrepit housing of Dsu standard, which was the same standard as Ludancun. The report indicated that the ratio of the cost to reinforce current buildings to the cost of demolition and reconstruction was 83.67%. Therefore, it suggested the application of the redevelopment approach instead of the reinforcement approach in Nanhuacun because the former can substantially address the issue of poor building quality.

After the promulgation of the test result, in May, the association invited several influential public media in Shenzhen to expose the dangerous building and terrible living conditions in Nanhuacun. The association intended to use public media to draw enormous public sympathy, which pressurized the local government for redevelopment approval (as an application of information and choice rules). As some owners noted,

> The housing in Nanhuacun have been identified as Dsu standard decrepit housing, which was the same with Ludancun. As Ludancun redevelopment commenced several years ago, why cannot the local government even set the redevelopment agenda for Nanhuacun? …The poor living quality results in the great inconvenience of our life. We have made tremendous contributions to Shenzhen's development since we came to this city. Now as Shenzhen has developed into a modern metropolis, it is extraordinary inappropriate that we are still living in the decrepit housing. ("The residents in Nanhuacun looking forward to redevelopment for more than 17 years", 2014)

The publicity strategy significantly pressurized the local government. Under pressure, the Shenzhen Housing and Construction Bureau and the Shenzhen Seismological Bureau sent a formal reply to the Nahuacun households in June 2014. Both bureaus recognized the decrepit housing of Dsu standard and promised to take instant actions to protect the safety of the affected residents. However, they indicated the residents had to refer to the Planning and Land Committee for redevelopment approval application. Residents then applied to the Planning and Land Committee but were informed that the residential redevelopment application was forbidden by the municipal government in Shenzhen. To seek more attention from the local government, some prestigious owners jointly issued a complaint letter to the municipal government for the agenda setting of Nanhuacun residential redevelopment in June 2014. However, there was not any response from the government after one month.

After realizing the passive attitude of the government, the resident association changed its strategy. It attempted to set up the property owner committee, a legal resident self-organized association, to deploy the equipped self-redevelopment power of the residents for an "owner-led" redevelopment (as the application of scope rules). With the owners committee under control, the resident association could not only apply to the redevelopment agenda on behalf of the affected residents; but also replace the Wanxia property management service provider, which they had a grudge with. According to Mr. Wu,

---

[4] The National Standard for Reliability Assessment of Civil Buildings promulgated by the State Council.

> After its formation, the resident association intended to transform the public space on the first-floor outside its apartment into a private garden, which would facilitate its publicity for redevelopment. However, it was obstructed by Wanxia, which triggered conflicts between two parties. In the building quality test process, the test institute sent some staffs on site to drill the wall. Wanxia and UCDC, as the property management service provider, impeded the processing of wall drilling. The resident association insisted on conducting the test, which escalated the conflict between the resident association and two property managers. (Interview of Mr. Wu, Jan-05-2017)

In October 2014, the resident association submitted an application to the work station for the election of property owner committee in Nanhuacun. After obtaining the agreement from its superior government agency—Nanyuan Street Office, the work station began to post posters and publicized the formal process in Nanhuacun. In November, the work station organized an on-site vote. Residents were required to vote for members of the property owner committee with their certificate of identification. However, the election process was obstructed by the conflict among the resident association, the residents and the work station. As residents noted,

> The staffs in the resident association organized more than 50 residents to block the entrance of the work station. With 300 RMB award, they encouraged us to vote the candidates working in resident association. Otherwise, it is impossible for us to enter work station to vote for other candidates. (Liu, 2014)

The staff in resident association argued that the work station was cheating in the election process:

> Only residents with their own identity cards could vote in accordance with the relevant regulations. However, the work station allowed one resident to vote several times with others' identity cards. This is corruption! The officials in work station supported candidates promoted by property providers and obstructed our staffs to be elected as owners committee members. This is because the work station has a corrupted relationship with the providers. Currently, both providers earned enormous money from parking services in Nanhuacun without inspection of the owners committee. They developed a good relationship and shared interests with the officials in the work station since we spotted, they banqueting the officials several times. As the newly established property owner committee will have authority to supervise the operation of the property management service providers, the current providers bribed the work station to cheat in the voting process…For the 300 RMB subsidy we provided, we just intended to reimburse their travel fee because many owners living outside Nanhuacun came back to vote. (Liu, 2014)

The close relationship with the property management service providers was denied by the chair of the work station. The chair proved that the two property management companies survived through the revenue of parking services because some households refused to pay the property management service out of the dissatisfaction to the chaotic community management (Interview of an official in work station, Jan-03-2017). However, Mr. Wu proved the close relationship between the work station and the property management service providers, and he denied that some residents refused to pay property management fees (Interview of Mr. Wu, Jan-05-2017).

Due to the conflict in the election process, the street office suspended the voting process. As the official in the street office indicated,

> It is acceptable that the resident association was willing to promote their staffs in the election process. However, the bribery to other residents and the interruption with the operation of the election process hurt the credibility of the election. (Interview of an official in Nanyuan Street office, Jan-12-2017)

The conflict in the property owner committee election drew the attention of the local media and further stirred the attention of Vanke. It soon publicly declared that it had never been involved in the Nanhuacun redevelopment by the end of 2014 (Interview of an official in Nanyuan Street office, Jan-12-2017). Subsequently, the resident association claimed that it only collected supports and invitation letters from the residents but did not hand it over to Vanke. Since then, increasing owners discovered the essence of the resident association. As a resident noted:

> The farce of the owners committee confused us. Why did they so eagerly promote their staff to be elected as members of the owners' committee? After the claim of Vanke, we eventually realise the resident association is a commercial agency organized by Dr. Shao. They planned to accomplish the residents' supports collection and redevelopment agenda through a so-called owner-led approach. On completion of that, they will sell Nanhuacun redevelopment as a package to Vanke or other private developers for profit. The generous compensation and relocation schemes were only promises by the association and Dr. Shao rather than Vanke. We were all fooled. (Interview of a resident in Nanhuacun, Jan-18-2017)

However, Mr. Wu, as a consultant of resident association, held a different opinion:

> The resident association is a commercial institute organized by Dr. Shao to seek fortune from Nanhuacun redevelopment. But this is acceptable if our residents could obtain the redevelopment approval with their assistance. (Interview of Mr. Wu, Jan-05-2017)

Since the claim of Vanke, the resident association had lost its credibility. Residents abandoned their dependence on the resident association for redevelopment agenda application and commenced petitioning to the local government by themselves. In March 2015, as a deputy to the National People's Congress (NPC) in Shenzhen, the owner Mr. Yang submitted residents' complaint to the Municipal Planning and Land Committee through the channel of NPC. At the meantime, some residents used their influence and submitted petitions to the provincial government (as the application of choice rule) (Interview of Mr. Wu, Jan-05-2017). They were all waiting for the response from the local government.

6.3.4 Government Involvement Arena

To respond to the community petition, the local government began to directly involve in Nanhuacun, which constructed the government involvement arena. On March 11, 2015, the Urban Redevelopment Office in Futian District officially rejected residents' request for Nanhuacun redevelopment agenda (as the application of choice rule) (Huang, 2015). It had been nine months after they applied to the district government. The report suggested that only the dilapidated residential areas which were identified as D-class would be approved for redevelopment in accordance with the

Decrepit Housing Appraisal Standard (DHAS hereafter). The Dsu standard buildings in Nanhuacun were inspected according to the National Standard for Reliability Assessment of Civil Buildings (NSRACB), which were not eligible for residential redevelopment application. For further confirmation of the building quality, the local government would conduct another test based on the DHAS. If only a small number of buildings were identified as D grade, the owners could apply for self-funded construction in the original site with the same building type. If large-scale dangerous buildings were identified as D grade, the owners could apply to the redevelopment office in the district government for redevelopment approval.

Out of dissatisfaction with the government's response, increasing residents petitioned to the municipal government and asked the latter to put pressure on the district government for Nanhuacun redevelopment (as the application of choice rule). To reduce grievance of the residents, in July 2015, the municipal government delegated the Municipal Housing and Construction Bureau to formally conduct a second test in accordance with the DHAS (as the application of scope rule). About a month later, on July 13, the MHCB posted posters in Nanhuacun and claimed that all four qualified institutes for building quality test in Shenzhen were invited to enhance the credibility, and the test would be accomplished within three months. However, many residents expressed their confusion and dissatisfaction to the announcement. According to Mr. Jiang,

> The local government built Ludancun and Nanhuacun in the same period with similar building technics. The construction of Ludancun was even two years later than Nanhuacun. The previous test is in accordance with the same standard as Ludancun. As the new Ludancun is under construction now, why do we still make efforts on the building quality test in Nanhuacun? (Huang, 2015)

Through a joint petition letter, the affected residents required the MHCB to execute the test according to the same standard with Ludancun. The MHCB replied that as the test was required by the Land and Planning Committee, they had no authority to shift the test standard, but the result of previous test would be submitted to the municipality for reference.

With three months passing, in October 2015, without receiving any notification, some residents came to the MHCB for inquiry (Shang, 2015). Officials replied that the test was suspended due to limited financial resources, but it would be rebooted in the next week. In November, 45 emotional property owners rushed into the municipal government to require the test result. However, the petition ended up with a verbal conflict between the residents and the officials. In February 2016, seven months after the announcement, there was still not any result from the government. To pressurize the local government, three residents guided many owners to the MHCB to show their opposition. To maintain social stability, the MHCB made a promise that the test would be accomplished in two months.

In April 2016, the MHCB published the result that the residential buildings in Nanhuacun were identified as decrepit housing of B class, which were not eligible for demolition and reconstruction. To improve the living environment in Nanhuacun, the municipal government would invest more than 13 million yuan in reinforcing the 20 most deteriorated buildings. As the affected households had to move out during the reinforcement, the government would reimburse the rental and relocation fee. Nevertheless, the test result triggered immense resistance from the affected households. They held the opinion that the government deliberately manipulated the result to obstruct the redevelopment agenda setting. As Mr. Wu noted,

> According to my observation, though the building quality is poor in Nanhuacun, the main structures will endure through time unless there is an earthquake or a strong typhoon. That's the genuine reason why the government dares change the Dsu standard decrepit housing to B class. The shift of the test standard is not convincing to our residents. If the cadres in the municipal government attached enough importance on Nanhuacun, the housing would be redeveloped even if it is a decrepit housing of A-class. Without a cadre in the local government caring about the living environment of Nanhuacun in the municipal government, the redevelopment will not be executed even if the housing is identified as D class. (Interview of Mr. Wu, Jan-05-2017)

For the test result, the officials in the street office explained,

> The NSRACB is like a medical examination for normal people whilst the DHAS is like a detection for patients of serious illness. Therefore, of course, the results are different. So, there is no 'black box operation'. The change of the test standard expressed the cautious attitude of the municipal government. Since the limitation of the superior policy – the property right law, the developer as well as the government is not able to utilize compulsory actions towards the last uncooperative residents. This results in severe project delay in the previously approved residential redevelopment cases. As there is not an efficient way to deal with the dilemma, the government has to postpone the residential redevelopment approval. (Interview of an official in Nanyuan Street office, Jan-12-2017)

The official working in the Urban Renewal Bureau explained,

> ...Ludancun is not an example of residential redevelopment in Shenzhen because of its extremely high administration cost to the local government. The size of Nanhuacun is almost two times to Ludancun. It is very difficult for the government to determine Nanhuacun redevelopment with the Ludancun mode. (Interview of YKL, Jan-12-2017)

After the promulgation of the result, the MHCB provided a reinforcement plan. However, the plan generated excessive resistance from the affected residents. As the residents noted,

> The government picked up the cheapest approach to reinforce the buildings in Nanhuacun due to its limited financial resources. But this method was only able to cover the exposed deteriorated steels rather than improve the stability of the main structure. In this case, the relocation for the reinforcement was only a waste of time for us. Before the International University Sports Federation, the municipal government spent enormous money in decorating the external walls in Nanhuacun to improve the artistic image of Shenzhen. When it comes to our living environment, the financial resources of the government become limited and constrained. Why not invest more money to improve residents living environment substantially? (Interview of the resident association, Jan-03-2017)

With the residents' fierce opposition, the government invested some capital in fixing the corridors and stairway, and the reinforcement for individual household failed to be carried out due to immense resistance (as the application of scope rule).

While the contradiction escalated between the residents and the government, the resident association invited a developer named Beijing Real Estate Company (BREC hereafter) to join in the preparation of Nanhuacun redevelopment in August 2016. The BREC offered a more generous compensation scheme than the resident association offered in 2013 (as the application of pay-off rule). According to the scheme, the DCR was 1:1.4. Every household was allocated with one parking spot. Any household which signed the contract would be awarded 120 thousand RMB. The rental subsidy and interior decoration compensation also were higher than the normal standard in the market. The affected residents who own a large apartment were even able to exchange for two small apartments at a discount price. Besides the compensation, the BREC promised to the affected residents that a reputable and competent developer would be invited if the Nanhuacun redevelopment was approved.

This attractive compensation scheme tempted some residents to sign contracts. However, none of the cooperative residents obtained the monetary award (Interview of an official in work station, Jan-03-2017). At the meantime, some residents found the BREC was never formally registered in the Chinese business system. With more solid evidence, some residents appealed to the district urban renewal bureau about the deceitful behaviours of the resident association. When asking his opinion about the resident association, the cadre replied,

> We know the background of the resident association. Dr. Shao bought an apartment as an office in the community, which is quite normal. We once tried to form a blacklist to warn the developers who continuously spent efforts in residential redevelopment. However, it is very difficult to define and further forbid its aggressive actions in the dilapidated residential area. What we do now is to invite them to our office and inform them to follow the policies the municipal government established. We wouldn't punish them unless they went too far. The main purpose of our government is to maintain social stability. (Interview of YKL, Jan-12-2017)

After three years of endeavours for redevelopment agenda application, both the resident association and the affected residents realized that it was impossible to obtain redevelopment approval through the solicitation of residents supports in Shenzhen (Tian, 2016). For the residents, they insisted on opposing the reinforcement approach and expected the government would eventually agree to the redevelopment by the Ludancun model. Based on their authoritative resources, residents kept petitioning to the municipal and district governments (as the application of choice rule).

6.3.5 Owner Motivation Arena

Since the residential redevelopment approval was prohibited by the local government in 2013, residents living in dilapidated areas like Nanhuacun kept petitioning to the local government for residential redevelopment. The municipal government established a Public–Private-Partnership (PPP) approach to facilitate residential redevelopment through uniting the three main parties: the state, the market and the residents. Residents were significantly motivated to initiate the redevelopment, which structured the owner motivation arena.

In January 2017, after the decentralization of redevelopment approval to the district government in Shenzhen, the district urban renewal bureau intended to investigate the living condition of Nanhuacun and planned to initiate residential redevelopment via the shanty area improvement approach (Interview of an official in Nanyuan Street office, Jan-12-2017). In May 2017, a draft policy promoting the Public–Private-Partnership (PPP) approach for the dilapidated residential redevelopment was published (Shenzhen Housing and Construction Bureau, 2017). The district government further commenced institutionalizing the approach by the new policy.

On June 16, 2017, some residents who got an internal message from the district government spread the news in the social media group that the district government would establish some temporary offices in the dilapidated residential area in Futian. These offices, as agencies for the local government, would guide the governmental department at grassroot level to solicit residents supports and set redevelopment agenda. They shared an internal document of the district government in the social media group. The document indicated that only the dilapidated residential areas would be approved which obtained supported by more than 90% of the residents within one week. If not, the district government would not consider the redevelopment in five years.

Once the message was released, some residents voluntarily strived to publicize the new model and motive more households through the social media group and an online forum (as the application of information rule). *"To improve our living conditions and leave a better living environment for our descendants, we encourage all the property owners to actively grasp this rare opportunity, publicize and inform each other, actively cooperate with the district government to commence the Nanhuacunredevelopment as soon as possible"* (Comments in the social media group, 2017).

In July 2017, the newly established Futian dilapidated residential redevelopment headquarter in District Urban Renewal Bureau spread out the information that they would investigate the living condition and solicit supports in the dilapidated residential areas which have been built for 20 years (Zheng & Ma, 2017). This information significantly encouraged the residents in Nanhuacun. At the meantime, the street office printed some documents for the residents to publicize the policy. To solve the questions of affected households, the work station held several meetings for the property owners and set up a dedicated reception in Nanhuacun. On August 4, the street office formally pronounced that from August 7, the property owners in Nanhuacun were able to submit their documentary support to the work station for the opinion solicitation (as the application of choice rule).

To tackle these problems, campaigns were commenced by a resident-founded volunteer team to publicize the new residential redevelopment policy (as the application of information and choice rules) (Zheng & Ma, 2017). Through house-to-house visit, they publicized the relevant regulations and requirement to the existing property owners. If the property has been rented out, they got in touch with the landlords with the assistance of the tenants. Through the publication of a working daily, the volunteer team updated the progress of the support solicitation in social media groups every day.

On September 29, within 53 days, they collected documentary supports from more than 2272 households, with another 24 verbal agreements (Zheng & Ma, 2017). While 147 cooperative households were preparing relevant documents, the rate of the cooperative households went beyond 90%. For the uncooperative households, some held a wait-and-see attitude in terms of being unfamiliar with the new redevelopment mechanism; some properties were in the mortgage and required time to deal with the formal procedure. However, for the properties that have been vacant for decades, it was very difficult to contact with the property owners and obtain their agreements. When asking about their opinion about the redevelopment, the volunteers were very optimistic. As the team head, Mr. Lin noted,

> Next, our work will focus on helping the 147 households to complement their relevant documents, winning more wait-and-see households and contacting the 'lost' property owners. The bottom line is the formal support from 90% of the households, and we anticipate achieving more than 95%. The more residents agree to the project, the more optimistic the government will be. The redevelopment, therefore, will get approved more rapidly. (Zheng & Ma, 2017)

6.4 Exogenous Influences and Their Impacts on Power Relations

The exogenous influences that have significant impacts on power relations in Nanhuacun redevelopment consist of three categories: the central policies, the local policies and community attribute.

The central policies influenced power relations from four perspectives: housing reform, property right reform, participatory redevelopment reform and the shanty area improvement institution. Through the housing reform in the 1990s, public housing was privatized into private ownership to each household, which nurtured their awareness of private property rights as structural power resources. In the first arena, the affected residents thus were empowered to oppose the unfair treatment of the local government and the involved developer. The property right reform in 2007 institutionalized the legitimacy of property right protection, which granted significant structural power resources to the property owners. The participatory reform in 2009 altered the power structure in the residential redevelopment projects. It required the local government to play a regulator to restrain its power in urban redevelopment, while empowered the affected residents to take part in the decision-making process. The shanty area improvement policy in 2013 maintained the local government's power to conduct compulsory acquisition, granted development right to the market sector and protected the participation of the affected residents.

The local planning policies also played an important role in influencing power relations. The urban renewal policy in 2009 authorized the development right as structural resources to both market sectors and the affected residents, for motivating their enthusiasm for urban redevelopment. Meanwhile, it regulates the necessity of the affected residents' participation in the decision-making process, which granted substantial structural resources to the residents for bargaining with the developer. In 2012, the municipal government prohibited the engagement of private developers to solicit residents' supports in dilapidated residential areas and reversed to the government-led approach. The development right of the developer was thus eliminated. In 2013, the inner document circulating in the local government forbade the approval of any categories of residential redevelopment, which withdrew the bargaining power and participating power from the affected residents. The limitation of DCR also constrained the structural resources of the affected residents to seek profit based on their property right. In 2016, The establishment of the PPP mechanism prevented the involvement of private developer in the agenda setting process of residential redevelopment and authorized the development right and housing expropriation power to the local government. It also granted bargaining power and participating power to the affected residents.

The influences of the community attribute on the local power relations have been identified, including the physical attribute, social composition, values and preference of the participants in the community. As a large-scale community with adjacency to Huaqiangbei, the commercial centre and other medical and educational facilities, the land of Nanhuacun has great potential value for redevelopment. The affected property owners thus hold solid allocative resources to bargain with the developer. Because of the inferior housing quality, many households chose to move out and lease their apartments for employers working in Huaqiangbei and surrounding areas. Due to the large size of the community, there are two property management service companies on site. The ambiguous division of these companies resulted in the conflicts and obstructed the foundation of the property owner committee. Without this official legitimate resident organization, the property owners in Nanhuacun are of a loose association. Some residents who were professors or senior engineers deployed their authoritative resources to pressurize the local government for redevelopment approval. In the multi-developer competition arena, the variations of participants' value and preference in a community resulted in different options towards the developers and thus obstructed the agenda setting of Nanhuacun redevelopment. It resulted in the involvement of the real estate agent, which significantly restructured power relations.

6.5 Endogenous Influences and Their Impacts on Power Relations

Four main stakeholders—the affected residents, the local government, the real estate agent and the developers—structured and restructured the power relations in Nanhuacun redevelopment through the configuration of resources and rules. Residents' power has its origin in three categories of resources, including the property ownership and the public participation mechanism of participatory redevelopment policy as structural resources; social relationship, reputation and professional skills as authoritative resources and premium location, large size and inferior living condition as allocative resources. In the first arena, based on the personal reputation and expertise, Mr. Wu led a small group of the affected residents to take mobilized actions to pressurize the local government for the redevelopment of their deteriorating building (as the application of choice rule). Considering that residents had a lower position compared to the local government, Mr. Wu adopted a collaborative strategy when negotiating with the local government (as the application of choice rule). His professional engineering skills empowered him to persuade the latter to adopt the demolition and reconstruction approach and abandon the reinforcement one. These rules served to achieve the residents' objective since a government-owned developer was arranged by the local government to conduct the redevelopment. Facing the defrauding behaviours of the developer, residents adopted mobilized actions to pressurize the developer (as the application of choice rule). This rule functioned since the developer consequently conducted the redevelopment. The participatory reform initiated the developer-led redevelopment in Nanhuacun in the second arena. Nevertheless, because of their preference variation, different residents were attracted and signed supporting documents with different developers, which prevented any developer from meeting the requirement of agenda setting. In the third arena, under the mobilization of the real estate agent, the affected residents were willing to deploy their self-redevelopment right to apply for the redevelopment approval (as the application of choice rule). The frequent usage of this rule did not play an effective role in changing the decision of the local government. In addition to the deceptive behaviours of the agent, the residents gave up the cooperation with the agent and reversed to mobilized actions to pressurize the local government for the redevelopment agenda (as the application of choice rule). Through collective actions, they pressurized the local government to conduct a building quality test for setting up the redevelopment approval (as the application of choice rule). However, the repetitive deployment of choice rules failed to work out and residents' power was counteracted by the power of the local government. After the establishment of the PPP approach, some residents formed a voluntary association as a community representative. They cooperated with the local government to publicize the new redevelopment mechanism and collect supports from the affected residents (as the application of information and choice rules). Despite with high frequency, these rules did not serve to support the redevelopment agenda setting so far.

The local government obtains its power from three categories of power resources, including structural resources—ruling position, land ownership, definition of regulations and policies and jurisdictions on land use and development intensity; allocative resources—capital investment; and authoritative resources—social relationships and control of government-owned developer. In the first arena, based on its authoritative and allocative resources, the local government possessed the decision-making power and proposed a reinforcement approach to address the petition of the affected residents (as the application of scope rule). This rule failed to come into force. The local government eventually resorted to a demolition and reconstruction approach and arranged a government-owned developer to improve the building quality (as the application of position rule). Facing with the poor performance of the approved developer-led redevelopment cases, the local government withdrew the decision-making power from the market sector and prohibited any types of residential redevelopment approval. It thus adopted an array of actions to dispel the efforts of the affected residents and the real estate agent (as the application of scope rule), including refusing to admit the supports collected by the agent and declining the result of building quality test conducted by the agent. It also postponed responding the petition letter of the affected residents to delay the project approval (as the application of scope rule). The repetitive deployment of these rules restrained the mobilization of residents' self-redevelopment right for the redevelopment agenda application. In the fourth arena, the local government manipulated to shift test standard to create obstacles for Nanhuacun redevelopment approval and promulgated the reinforcement approach to improve living condition (as the application of scope rule), which triggered intensified resistance from the affected residents. Since the establishment of the PPP mechanism, the local government delegated the grassroot governmental institute to collect supports from the affected residents for redevelopment approval (as the application of choice rule), which failed to apply for the agenda so far.

Without any structural resources, the real estate agent procures its power primarily from its financial strength as allocative resources; and reputation and social relationship as authoritative resources. In this case, the agent attempted to play a power broker of the affected residents to deploy the latter's self-redevelopment power. It organized the resident association and adopted an array of publicity strategies to build up its reputation (as the application of choice rule); and also provided a highly beneficial compensation scheme for the affected residents (as the application of payoff rule). These rules served to achieve its objective to obtain supports from the residents. Based on the collected supports, the real estate agent deemed itself had a higher position than the local government so that it began to adopt strategies to pressurize the local government for the redevelopment approval. It invested to conduct a building quality test to provide rationality for the redevelopment and invited public media to expose the inferior living condition to pressure the local government (as the application of information and choice rules). In addition, it intended to organize the POC to enhance the residents' self-redevelopment right (as the application of scope rule). However, these rules failed to influence the decision of the local government. The declaration of Vanke significantly hit its reputation and destroyed its well-maintained social relationship with the affected residents, which reduced its authoritative resources. Without these resources, the agent was thus marginalized in power relations. Being reluctant to quit from Nanhuacun, in the fourth arena, it invited another real estate company to offer higher compensation terms for support collection (as the application of pay-off rule). The rule failed to take effect because of the absence of power resources of the new company. The real estate agent eventually quit when the PPP mechanism was established.

Two categories of developers were identified in the "owner-led" redevelopment process. The first one refers to the government-led developer, and its power was originated mainly in the decentralized land development right and the close relationship with the local government. As a power broker of the local government, it endeavoured to relocate the residents and improve the building quality by reinforcement approach (as the application of choice rule). This strategy triggered substantial resistance, which forced it to resort to the redevelopment approach. Before the reconstruction was accomplished, it took opportunistic actions for obtaining reimburse from the affected residents (as the application of pay-off rule), which then counteracted by the resident leader who had the close relationship with the manager. The other category of developer signifies the marketized developers which joined the redevelopment in the second arena. Obtaining their power from allocation resources including expertise, capital and reputation, they adopted strategies to raise compensation terms and collect supports from the affected residents (as the application of choice rules). However, the intense competition obstructed any of them to obtain supports from the majority of the affected residents and thus the redevelopment right, which failed in agenda setting.

6.6 Outcome and Evaluation

6.6.1 Outcome

Redevelopment agenda has not been approved in Nanhuacun although involved stakeholders, such as the local government, private developers, real estate agent and involved residents, have been taking enormous endeavours for more than nine years. In 2017, the Futian District Government commenced to experiment with the PPP approach in several dilapidated residential redevelopment projects; however, Nanhuacun failed to be listed (Futian government, 2017). The decision intensified the resistance of the affected residents and triggered more petitions to the local government. In the community, residents are still suffering by living in inferior environment and the volunteer team was devoted to the cooperation with the government for support solicitation. The real estate agent led by Dr. Shao closed the resident association and quit the project in 2018.

6.6.2 *Equity Evaluation of the Outcome*

As the redevelopment project has not been approved in Nanhuacun, there is no anticipating benefit generating from the process. Each individual took rational action but aggregated to produce socially irrational outcomes. For the residents, except a few residents obtained newly constructed housing in 2003, most residents have been suffering from the poor living condition for almost 20 years. They, on the one hand, devoted efforts to cooperate with diversified actors, such as the private developer, the real estate agent and the local government, to apply the redevelopment agenda in different arenas. On the other hand, they utilized their resources to adopt mobilized actions such as petition to governments at multiple levels for project application. However, none of these efforts served to succeed.

In this case, the government invested considerable funding to initiate the second test with four eligible institutions and to conduct the reinforcement program to strengthen the structure of decrepit buildings. The administrative cost of the government for dealing with residents' repeated petitions is supposed to be immense. The benefit of the government in the process, whereas, is none.

After the promulgation of market-oriented mechanism, private developers engaged in Nanhuacun and invested in collecting residents' supports. Because of the fierce competition among these involved developers, none of them have succeeded in solicitating agreements from more than two in three of the residents and thus became the candidate. Neither did they profit from the involvement. The real estate agent injected a lot of manpower, resources and capital investment into the project, regarding the cost for hiring consultants and staffs for the resident association, expenditure of maintaining a good relationship with residents and investment for the first building quality test. Nevertheless, the agent did not get any return from the process. Even if Nanhuacun redevelopment succeeded in the future, the previously engaged developers and the agent would not share the interests. The interest redistribution in Nanhuacun redevelopment is evidently inconsistent with the equity principle that those who derive greater benefits are expected to pay more (Imperial, 1999).

6.6.3 *Institutional Adaptation of the New Power Structure*

In Nanhuacun, different stakeholders manifest different levels of adaptability to the power dynamics changes. Before the establishment of the market-oriented redevelopment policy, despite a dominant role, the local government exhibited a high level of adaptability to power structure in urban redevelopment. Residents with expertise persuaded the local government to adopt a demolition and construction approach to improve living condition in Nanhuacun. To pacify social resistance, the municipal government adopted a practical strategy and gave access to private developer to conduct redevelopment on a small scale to address the terrible building quality. After the promulgation of the market-driven approach to urban redevelopment, the government transformed to be a regulator with limited direct intervention in urban redevelopment. However, as the market-oriented mechanism generated severe issues in practice, the government abolished the market mechanism and suspended the further approval of residential redevelopment. Since then, the local government became decreasingly adaptable by resorting various strategies to suppress residents' requirement.

The real estate agent showed a notable level of adaptability to the power dynamics change. At first, the agent attempted to mobilize the affected residents to commence an owner-led redevelopment to avoid the prohibitive regulation of the private developer's involvement in residential redevelopment. As the strategy was unsuccessful in achieving the goal, the agent initiated a building quality test to draw the public attention and pressurize the local government for project approval. Since the new strategy did not serve to make a difference, it further invested in the selection of property owner committee to resort to more structural power resources for achieving its aim. In the government-led arena, when the contradiction between the government and the residents was escalated after the second building test, being reluctant to quit, it invited another private company to collect supports from the affected residents to initiate owner-led redevelopment.

Residents adopted strategies in the different power structure to achieve their ends, manifesting outstanding adaptability. Without power to bargain for redevelopment in the owner-led arena, residents held a cooperative attitude towards the local government and persuaded the latter with expertise and knowledge for the permission of redevelopment in small scale. Once they were delegated power to join in the decision-making process in the multi-developer competition arena, they made a comparison among different developers and held a wait-and-see attitude for higher compensation. After the suspension of the residential redevelopment approval, residents kept petitioning to the government and then cooperated with the resident association to pressurize the government through building quality test. Once the government publicized the PPP mechanism, residents turn to a cooperative strategy and formed voluntary team to assist with the support solicitation.

6.6.4 Transaction Cost and Project Delay

Transaction costs incurred in the Nanhuacun project for applying for redevelopment approval from the local government. Field evidence shows that transaction costs in Nanhuacun are associated with stakeholders' behaviours and power relations.

Power asymmetry characterized the agency guidance arena and the government involvement arena. In the former arena, the resident association and the real estate agent had ample allocative and authoritative resources while the grassroot government institutions had much fewer resources. The resident association obtained supports from the majority of residents and more authoritative resources skewed to the agent. The resident association deems itself to became more powerful and had a higher position compared to the grassroot government institutions. It subsequently adopted opportunistic behaviours in the selection of property owner committee and adopted antagonistic strategies towards the local government. This resulted in resistance from the latter and created obstacles to agenda setting, adding up to the coordination cost of both parties. Power asymmetry occurring in the government involvement arena delayed the redevelopment application and escalated transaction cost. The local government had stronger structural resources to decide on redevelopment approval while the affected residents had seeming self-redevelopment power for redevelopment application. Despite various endeavours of the residents, the government changed the building quality test standard and escalated the requirement of redevelopment approval. The overwhelming power of the local government significantly delayed the redevelopment process and thus amounted to the transaction cost.

6.7 Discussion

This section examines the advantages and disadvantages of owner-led power relations in delivering an urban redevelopment project. Market-oriented reforms aim to allow affected residents to initiate redevelopment and thereby capture a larger share of the benefits. However, due to high transaction costs and weak institutional arrangements, Shenzhen's owner-led approach has proven largely ineffective in facilitating residential redevelopment. Poor institutional design imposed substantial time and economic costs on all involved actors without generating meaningful benefits. Four institutional obstacles have been identified in the owner-led redevelopment process.

First, in an effort to share more benefits with affected residents, the 2009 market-oriented approach was intended to incentivize developers to engage directly with residents to secure their support for redevelopment projects. In practice, however, multiple developers competed within the same residential area, investing considerable resources—such as publicity campaigns, financial incentives and compensation offers—to secure redevelopment rights. These competitive strategies fragmented residents' support across different developers, undermining effective agenda setting in Nanhuacun.

Second, frequent shifts in residential redevelopment mechanisms created policy ambiguity, which confused the stakeholders. After recognizing the shortcomings of developer-led projects, in 2012, the local government prohibited private developers from participating in future residential redevelopment cases and reverted to a government-led approach for dilapidated residential areas. However, this policy did not eliminate the self-redevelopment rights of affected residents, resulting in unclear institutional guidance. Consequently, both residents and market actors attempted to initiate owner-led redevelopment, leading to substantial waste of social resources.

Third, the absence of a self-organized community association posed significant barriers to residential redevelopment. As a self-organized body, the property owners' committee is expected to play a central role in coordinating community affairs, given its close linkage to the interests of residents (Chao, 2003). However, in Nanhuacun, self-redevelopment organizations such as a property owners' committee do not exist. Consequently, market actors, including real estate agents, became involved to initiate "owner-led" redevelopment, seeking to capture substantial profits. In Nanhuacun, the real estate agent both engaged in opportunistic behaviours that increased the social costs of the project and generated conflicts among itself, the grassroots government, and affected residents. Together, these factors significantly hindered the redevelopment process.

Fourth, the local government, leveraging its dominant power in the owner-led redevelopment process, acted in politically expedient ways that created additional barriers to residential redevelopment. To prevent the approval of the Nanhuacun redevelopment, the local government altered the standards for building quality assessments, which significantly undermined its credibility among affected residents. At the same time, in an attempt to address issues arising from the market-driven approach, the Shenzhen municipal government promoted reinforcement measures to improve the poor living conditions in dilapidated residential areas. While this strategy offered a temporary and evasive solution, it entailed substantial costs. Due to strong resistance, the reinforcement approach was rarely implemented in Shenzhen. These politically expedient measures failed to meaningfully improve residents' living conditions while generating significant social costs.

6.8 Summary

This chapter focuses on analysing the power relations in the owner-led participatory redevelopment process, as summarized in Table 6.1. Although participatory redevelopment policies theoretically grant affected residents the authority to initiate redevelopment, in Shenzhen the local government has been reluctant to fully implement the owner-led mechanism. In practice, the local government exercises dominant power and constrains the agency of other stakeholders. It oversees the entire decision-making process, including the confirmation of the redevelopment approach, building quality assessments and project approval. The authority of private developers is limited through regulations that prohibit their involvement, while residents, without formal authorization for self-redevelopment, cannot obtain project approval to proceed independently. Real estate agents attempting to act as intermediaries on behalf of residents are similarly unable to leverage influence over the decision-making process. As a result of the local government's authoritarian control, power relations in this case remain conflictual and contradictory, failing to evolve into a collaborative and inclusive dynamic.

Table 6.1 Summary of power relations in the owner-led case

Actor	Rules employed	Resources mobilized
Phase 1: Owner-led Arena (2000–2004)		
Resident leader (Mr. Wu)	• Aggregation rule • Choice rule • Information rule	• Authoritative resources
Property owners	• Aggregation rule	• Authoritative resources
Municipal government	• Position rule	• Structural resources
SOE developer (UCDC)	• Choice rule • Pay-off rule	• Allocative resources
Equity: Relatively high Adaptability: High Transaction Costs: Moderate		
Phase 2: Multi-developer Competition Arena (2009–2012)		
Private Developers	• Choice rule • Pay-off rule • Information rule	• Structural resources • Allocative resources • Authoritative resources
Residents	• Choice rule • Scope rule	• Structural resources • Authoritative resources
Equity: Low Adaptability: Low Transaction Costs: High		
Phase 3: Resident Association Arena (2013–2014)		
Real Estate Agent (Dr. Shao)	• Aggregation rule • Scope rule • Choice rule • Information rule	• Allocative resources • Authoritative resources
Residents	• Position rule • Choice rule	• Structural resources • Authoritative resources
Equity: Low Adaptability: Moderate Transaction Costs: High		
Phase 4: Government Involvement Arena (2015–2016)		
Municipal & District Government	• Choice rule • Scope rule • Boundary rule	• Structural resources • Authoritative resources
Residents	• Choice rule • Information Rule	• Authoritative resources • Structural resources
Equity: Mixed Adaptability: Moderate Transaction Costs: High		
Phase 5: Owner Motivation Arena (2017–)		
Municipal & District Government	• Boundary rule • Position rule	• Structural resources • Authoritative resources

(continued)

Table 6.1 (continued)

Actor	Rules employed	Resources mobilized
Residents	• Choice rule • Information rule	• Structural resources • Authoritative resources
Equity: Low Adaptability: Low Transaction Costs: High		

References

Chao, C. (2003). Democratic electoral reform for communities in China.

Futian government. (2017, September 4). New residential redevelopment model has commenced in Futian district. *SZ.GOV.CN*. http://www.sz.gov.cn/cn/xxgk/zfxxgj/gqdt/201709/t20170904_8419609.htm

Huang, C. (2015, March 13). The building quality test has to follow the relevant regulation in Nanhuacun. http://www.zhfutou.com/news/content/2015-03/13/content_11298432.htm

Imperial, M. T. (1999). Institutional analysis and ecosystem-based management: The institutional analysis and development framework. *Environmental Management, 24*(4), 449–465.

Liu, Y. (2014, November 14). The Nanhuacun redevelopment is about to commence. Do anyone have secret ties with developers?

Shang, J. (2015, November 3). 2566 households in Nanhuacun living in decrepit housing. *OEEEE.com.*

Shenzhen Housing and Construction Bureau. (2017). *Several measures on accelerating the Shanty areas improvement (Draft).*

SzhomeBBS. (2010, March 11). Nanhuacun redevelopment is about to commence. Which developer do you prefer? http://bbs.szhome.com/30-25942-detail-92625752.html

SzhomeBBS. (2013, December 19). The substantial implementation of Nanhuacun redevelopment has began.

Tian, C. (2016, January 14). Dozens of housing were identified as decrepit housing but the redevelopment is in stall. *Shenzhen News.* http://dc.sznews.com/content/2016-01/14/content_12735295.htm

Zheng, Y., & Ma, L. (2017, October 01). The largest project of shanty area improvement will commence soon with more than 90% of supports in 53 days. *Shenzhen Sina Real Estate.* http://k.sina.com.cn/article_1069205631_3fbac87f020004qrq.html

Chapter 7
Discussion

7.1 Introduction

This chapter discusses the empirical findings in the context of literature. Section 7.1 discusses the changes in power relations in urban redevelopment in China. Section 7.2 concerns exogenous influences, and Sect. 7.3 concentrates on endogenous influences. Section 7.4 connects the discussions on power, power relations and institutions. Section 7.5 summarizes.

7.1.1 Discussion on Power Relations

This research provides an alternative interpretation of power relations among key stakeholders in participatory redevelopment, complementing the existing studies reviewed in Sect. 1.2 of Chapter 1.

Government-Led Redevelopment

Relevant literature (e.g. Han & Wang, 2003; Shin, 2009) on the government-led projects in the marketed-oriented reform period suggests that local governments played a central and decisive role in the redevelopment process. Municipal authorities exercised substantial power over key aspects of redevelopment, including project approval, the formulation of compensation schemes, land clearance and land auctioning (Han & Wang, 2003). To consolidate and extend their control over the redevelopment process, local governments often transformed their affiliated entities into newly established development corporations that operated as local agents (Shin, 2009). These local agents were primarily driven by profit extraction rather than concerns for social stability.

X. Li, *Participatory Redevelopment as Contested Arenas in China*,
https://doi.org/10.1007/978-981-95-9394-1_7

The power relations observed in the government-led participatory redevelopment project examined in this research, however, diverge from these established accounts. Although the local government continues to play a dominant role by dictating project approval, compensation arrangements, land assembly and demolition, it has also established an on-site office to negotiate directly with affected residents and to secure their support. This on-site office performs a role distinct from that of the local development agents identified in earlier studies, as it does not operate solely with the objective of maximizing profits from residential redevelopment. Instead, both the local government and the on-site office have placed increasing emphasis on residents' welfare and the maintenance of social stability.

To expedite redevelopment during the market-oriented reform period, local governments commonly formed pro-growth coalitions with private developers possessing substantial financial and technical capacities. In such arrangements, developers typically initiated redevelopment projects and unconditionally complied with government decisions concerning land assembly, relocation and construction (Han & Wang, 2003; Zhang, 2002). In contrast, the case examined in this research shows that the local government only invites a private developer to take charge of reconstruction after the completion of land assembly. The developer, no longer operating as a coalition partner of the local government, assumes a largely contractual role and possesses limited capacity to influence decision-making.

During the market-oriented reform period, local communities generally occupied a marginal position in redevelopment processes and lacked meaningful power to influence decision-making (Shin, 2009; Yang & Chang, 2007). In this research, however, participatory redevelopment policies are found to have conferred substantial power on affected residents. Rather than being excluded, residents are actively involved in the decision-making process and are able to counterbalance governmental authority by adopting diverse strategies to negotiate for a greater share of land value increments.

Existing studies on government-led redevelopment in the participatory period suggest that the previously overwhelming power of local governments has gradually receded, as forceful strategies and unilateral actions have become less effective in implementing redevelopment projects (e.g. Zhai & Ng, 2013; Zhang et al., 2016). Affected residents have been increasingly empowered to bargain with local governments over compensation arrangements, requiring the local state to secure the support of a majority of residents in order to complete land assembly. As a result, urban redevelopment has become more inclusive and collaborative. The findings from the government-led participatory redevelopment project examined in this research largely support these arguments. However, unlike government-led projects in other cities, affected residents in this case exhibit stronger capacity to influence governmental decisions and to resist perceived unfair treatment. This heightened resident power can be attributed to Shenzhen's formal public participation mechanisms, which institutionally grant residents greater decision-making authority than those in cities such as Beijing and Xi'an.

Developer-Led Redevelopment

Existing studies on developer-led redevelopment during the marketization period indicate that international developers typically dominated the decision-making process, giving rise to a developer-led mode of redevelopment (He & Wu, 2005; Yang & Chang, 2007). Through the establishment of local agents and the promulgation of preferential policies, local governments primarily assumed a facilitating role in land assembly and site clearance for international developers. In contrast, affected residents possessed limited bargaining power and had minimal capacity to influence the decisions of either local governments or international developers.

Studies on the developer-led redevelopment in the participatory period reveal that the developers continued to dominate the redevelopment process across key stages, including the formulation of redevelopment plans and designs, project construction and project sales (Cheng, 2012; Li & Li, 2011). Compared with earlier periods, community actors—comprising affected residents and involved villagers—possessed greater bargaining power to negotiate with resource-rich actors such as local governments and developers in pursuit of fairer compensation. In this context, local governments primarily assumed the roles of regulator and facilitator, coordinating relationships and mediating interest distribution between developers and involved villagers (Li & Li, 2011; Wu, 2018). The collaborative relations established among these three stakeholders were considered conducive to the successful completion of redevelopment projects.

The power relations in developer-led participatory redevelopment identified in this research differ markedly from those reported in the existing literature. In this case, the private developer is authorized to initiate and implement the redevelopment project through direct negotiations with affected residents. However, owing to constraints imposed by central-level policies, local government intervention is substantially limited. Without formal endorsement from the local government, the developer lacks sufficient authority to counterbalance residents' power. The participatory reform has substantially empowered residents by granting each household the ability to suspend the redevelopment project through refusal to sign the consent agreement. As a result, residents—rather than developers or local governments—unexpectedly occupy the most influential position in the decision-making process. Residents are further differentiated into two groups: cooperative residents who support the developer and uncooperative residents who contest the redevelopment.

Cooperative residents who have signed the agreement transfer their property rights to the developer and are subsequently marginalized in the decision-making process. In response, they attempt to form a coalition with the developer to consolidate power and exert pressure on uncooperative residents. Conversely, uncooperative residents deploy their veto power by refusing to sign the agreement, with some engaging in speculative behaviour to maximize individual gains. These interactions generate conflictual and often chaotic relations among the three parties. Meanwhile, the local government remains largely confined to policymaking and project approval, possessing limited capacity to regulate the behaviours of the contracting parties or to resolve the emerging conflicts.

Owner-Led Redevelopment

Existing studies on power relations in owner-led redevelopment have primarily focused on urban village redevelopment (Cheng, 2012; Chung, 2013; Zhou, 2014). In these cases, villagers are authorized by local governments with self-redevelopment rights, enabling them to initiate owner-led redevelopment projects in urban villages. The village collective, acting as the formal representative of villagers, assumes a leading role in the redevelopment process. On the one hand, it negotiates with developers to secure higher compensation for both the collective and individual villagers; on the other hand, it seeks to persuade all villagers to sign redevelopment agreements in order to facilitate project implementation. Throughout this process, local governments deploy administrative power to mediate conflicts between the village collective and uncooperative villagers, while developers primarily exercise financial power to pursue profits from redevelopment.

In contrast, the power relations observed in owner-led participatory redevelopment in this research differ fundamentally from those identified in urban village contexts. Here, the local government retains a decisive and dominant position in decision-making. It possesses the authority to prevent developers from initiating residential redevelopment projects and to constrain both real estate agents and affected residents from exercising their nominal self-redevelopment rights. Through the promulgation of new redevelopment mechanisms, the local government is also able to restructure relationships among the involved stakeholders.

Contrary to policy expectations, residents' power is not substantially enhanced in owner-led redevelopment of dilapidated residential areas, particularly when compared with villagers authorized with self-redevelopment rights in urban village redevelopment. Although policies formally grant residents the right to initiate redevelopment, self-collected redevelopment support is not recognized by local governments in practice. This effectively renders residents' self-redevelopment power non-existent, preventing them from meaningful participation in decision-making processes. Moreover, a new stakeholder—the real estate agent—emerges in owner-led residential redevelopment. In the absence of a collective representative comparable to the village collective, real estate agents attempt to act as power brokers on behalf of affected residents by mobilizing support and applying for redevelopment approval. However, lacking a legally recognized identity and formal authority, their interventions fail to gain acknowledgement from local governments.

Power Relations in Urban Redevelopment

Existing studies in the participatory redevelopment period suggest that the affected residents who were previously excluded have become increasingly involved in the decision-making process in urban redevelopment (e.g. Chung, 2013; Zhai & Ng, 2013; Zhang et al., 2016). In these accounts, local governments primarily assume facilitating and regulatory roles, as state power is constrained by the rising influence of community actors (Li & Liu, 2018; B. Li et al., 2019). Changes in redevelopment mechanisms are argued to have enabled the three main stakeholders—the state, developers and communities—to act as co-protagonists in urban redevelopment, leading to more inclusive and collaborative power relations (Zhang et al., 2016).

However, the power relations observed in participatory redevelopment in this research diverge substantially from those depicted in the existing literature. Four key stakeholders are identified: the local government, affected residents, developers and real estate agents. Although residents have indeed gained greater bargaining power and play a more prominent role in decision-making—consistent with the participatory redevelopment narrative—the process of empowerment is highly differentiated and fragmented. Households whose properties are included in redevelopment projects, whether under government-led or developer-led approaches, are substantially empowered to negotiate with resource-rich actors (i.e. local governments or developers) for improved compensation. In contrast, households whose properties are not scheduled for redevelopment are excluded from both empowerment and decision-making. Moreover, residents in dilapidated residential areas are not effectively authorized with self-redevelopment rights.

In Shenzhen's participatory redevelopment, local government power has not been constrained by community power. Instead, the local state continues to occupy a dominant and decisive position in the decision-making process, far exceeding the conventional roles of facilitator or regulator (Wu, 2018; Wu et al., 2022). In owner-led redevelopment, the local government deploys a range of strategies to override the power deployment of other actors, including developers, residents and real estate agents. In government-led redevelopment, despite resistance from dissenting residents, the local government retains the authority to compulsorily acquire land to advance redevelopment. In developer-led redevelopment, although direct intervention is institutionally restricted, the local government is able to abolish market-oriented mechanisms for future residential redevelopment in order to maintain control over localized redevelopment projects.

Compared with the market-oriented reform period, developers possess relatively limited power in participatory redevelopment (Chen et al., 2022). In government-led projects, developer involvement is delayed, and developers are largely confined to the role of builders with minimal influence over decision-making. In developer-led projects, developers lack the authority to counter residents' resistance without local government endorsement. In owner-led redevelopment, developers invest substantial resources to mobilize residents' support but fail to obtain redevelopment approval. Overall, developers rarely occupy a dominant position in participatory redevelopment, in contrast to their roles during the market-oriented reform period or in urban village redevelopment under participatory policies.

The real estate agent emerges as a new actor within the market sector, attempting to function as a power broker by mobilising residents' nominal self-redevelopment rights. Following administrative conflicts between different levels of government, the Shenzhen municipal government abolished the developer-led redevelopment mechanism and suspended approval of dilapidated residential redevelopment projects. Consequently, the owner-led approach became the only remaining pathway for residents in dilapidated residential areas to initiate redevelopment. In the absence of a collective representative capable of coordinating fragmented residents, real estate agents established community associations and sought to act as intermediaries in applying for owner-led redevelopment. However, because owner-led residential redevelopment was never approved by the local state, residents' self-redevelopment rights remained ineffective in practice, rendering the power-brokering role of real estate agents largely invalid.

Other stakeholders—including the central government, banking institutions and demolition companies—also participate in participatory redevelopment, albeit in more limited ways. Unlike its dominant role during the socialist period, the central government rarely intervenes directly in local decision-making, instead exerting influence primarily through the promulgation of national policies. Compared with cities such as Xi'an (Zhai & Ng, 2013) and Beijing (Zhang et al., 2016), direct central-state intervention in participatory redevelopment is particularly rare in Shenzhen. The role of banking institutions has also diminished. During the market-oriented reform period, developers relied heavily on bank loans to fulfil financial obligations, enabling banks to exert influence over developers' decisions (Han & Wang, 2003; Wu, 2004). Following participatory reform, land transfers increasingly occur through private negotiation between developers and existing land users, reducing upfront financial requirements and, consequently, developers' reliance on bank financing. As a result, banks no longer play a decisive role in participatory redevelopment.

Similarly, the role of professional demolition companies has shifted. While earlier studies observed that demolition companies employed coercive strategies to expedite relocation under developer pressure (Wu, 2004), this research finds that resident empowerment constrains such practices. Although developers continue to hire demolition companies, these firms are increasingly confined to providing technical demolition services and are unable to accelerate relocation processes through forceful means, particularly in government-led projects.

While existing studies characterize participatory redevelopment as fostering more inclusive and collaborative power relations (Chung, 2013; Zhang et al., 2016), Shenzhen's experience presents a contrasting picture. Expanded resident participation has increased the number of stakeholders involved, but the absence of collective organizations capable of internalizing heterogeneity has resulted in fragmented and dispersed negotiations. Without well-designed institutional arrangements, participatory redevelopment has become an arena of intensified conflict rather than cooperation. Stakeholders possess distinct resources and pursue divergent strategies to maximise individual interests, producing division and confrontation more frequently than unity. Neither stable growth coalitions nor fixed urban regimes have emerged to facilitate redevelopment, rendering power relations increasingly conflictual and chaotic.

The occurrence of the real estate agent to some extent demonstrates the empowerment of the affected residents. According to current studies, only powerful actors are able to have agents as a local representative to influence the decision-making process. For example, in the socialist period, the central government extended their control over land development at a local level by decentralizing the decision-making power to some local agents (Yeh & Wu, 1996). In the marketed-oriented reform period, the local government or the SOEs once set up government-owned organizations or SOEs-affiliated developers as local agents to take control of urban redevelopment (Zhu, 1999). In this research, the real estate agent acts as a power broker in dealing with the affected residents in the decision-making process. This demonstrates the fact that the residents possess an increasing amount of power which cannot be ignored in participatory redevelopment.

These reconfigured power relations have significant implications for both the process and outcomes of participatory redevelopment. Similar to its entrepreneurial role during the market-oriented reform period (He & Wu, 2009), the state continues to promote redevelopment to sustain urban and economic growth. However, in participatory redevelopment, local governments employ more sophisticated strategies to extract benefits. Rather than relying solely on land value capture through taxation and land-use fees, governments obtain redeveloped land, infrastructure and public housing in kind and often free of charge. Under the current fiscal system, land-related revenues remain crucial to local finances, compelling local governments to pursue growth-oriented redevelopment strategies. Meanwhile, the proliferation of stakeholders complicates property rights and interest configurations, increasing the transaction costs of redevelopment. Developers must therefore invest substantial human and financial resources in negotiating with residents, understanding heterogeneous demands and devising tailored strategies to secure consent, significantly raising the overall cost and complexity of participatory redevelopment.

7.1.2 Discussion on Exogenous Influences

The exogenous influences on power relations in participatory redevelopment project are from three sources. They are central policies, local land use and planning policies and the physical and social attributes of the community. Through the central policies, the central government promotes a power decentralization process to empower the residents and constrain the power of the local government. The local government, through local land use and planning policies, strives to remain its control over urban redevelopment projects and constrain the empowerment of the affected residents. These local efforts demonstrate the countermovement of power decentralization. The community attributes, in terms of physical and social influences, determine which redevelopment approach is suitable to apply to one participatory redevelopment project. The applied redevelopment approach has a great impact on the power structure in the specific project.

Central Policies

An array of central policies launches a new era of participatory redevelopment to address the institutional obstacles of the traditional government-monopolized redevelopment mechanism, in which power relations among the main stakeholders are restructured. However, current studies primarily concentrate on the impact of the *San Jiu Gai Zao* policy. This research argues that a range of central policies forms an institutional reform to restructure the power relations in urban redevelopment from several perspectives, including constraint on rural land transfer, institutionalization of property right, new land transfer approach and public participation mechanism.

The central government establishes policies to restrict the rural land conversion, which limits local governments' power to appropriate agricultural land in periphery urban area on one hand. On the other, these policies accent the importance of urban redevelopment practices in Chinese cities, especially the megacities where acute land scarcity emerges to hinder further urban expansion.

The property right reform formally legitimatizes and institutionalizes the protection of private property rights. It grants strong structural resources for the property owners and empowers them not only to oppose compensation terms they deem unfair, but also to possess and benefit from the private property ownership. More importantly, it conceptualizes these resources into cultural and moral notions which subsequently shape residents' code of conduct (Abramson, 2011). Residents commence holding a firm belief that they are empowered to adopt resisting strategies and actions when facing unfair treatment and enforced behaviours (Po, 2011). Since the government-dominated land transfer mechanism conflicts with the property right protection, mounting opposition and protestation from the community emerge, which creates barriers to land assembly implementation (Mertha, 2009).

To facilitate urban redevelopment, the *San Jiu Gai Zao* policy permits the direct land use transfer between the existing land users and potential land users, which removes the supply side constraints for the developer in urban redevelopment (Ye, 2011). The policy regulates the local government to play a conductive and regulative role in specific urban redevelopment cases and empowers the affected property owners to take part in the decision-making process by directly negotiating with the involved developer. In this regard, the policy grants development right to developers and self-redevelopment rights to existing land users, promoting the alliance formation among the involved stakeholders (the state, market sectors and communities) (Li & Liu, 2018; Tian & Yao, 2018).

The abolition of the demolition permit institution in 2011 further restrains the intervening power of the state in the commercial redevelopment projects. Through the public participation mechanism, the policy enables the residents to join the decision-making process. However, the constraints of the local government intervention in commercial redevelopment cases extremely empowers the affected residents in the market-oriented cases in Shenzhen.

The central policies, on one hand, constrain the direct involvement of the local government and encourage the latter to play a conductive and regulatory role in participatory redevelopment. On the other hand, they grant structural resources to the affected residents through the institutionalization of property rights and public participation mechanism and to developers through direct land transfer mechanism. These central forces attest that through power decentralization to the affected residents, the central government aims to abolish the traditional government-monopolized redevelopment mechanism to pacify social resistance and facilitate urban redevelopment.

Local Land Use and Planning Policies

Existing literature fails to pay enough attention to the influences of local land use and planning policies on power relations. This research argues that as a response to the central policies, local land use and planning policies materialize and localize these central forces to impact power relations at local level.

In response to the participatory redevelopment reform at central level, in 2009, the Shenzhen municipal government promulgated two milestone policies to establish its local redevelopment mechanisms. Through the "Measures for urban renewal in Shenzhen", the local government set up a market-oriented redevelopment mechanism. The local government merely focuses on policymaking and project approval; the developer is authorized development right to directly negotiate with the property owners for the land use transfer through the contract, forming the developer-led mechanism. In the meantime, property owners are authorized self-redevelopment rights to initiate owner-led redevelopment.

However, the market mechanism triggered vicious competition between different developers and resulted in chaos and conflicts in some residential areas, which incited considerable social resistance, as mentioned in Chapters 5 and 6. In addition, in 2011, the abolition of the demolition permit scheme at central level significantly constrained the intervening power of the local government in the commercial redevelopment projects. As a local response, in the next year, Shenzhen established a new policy which regulated that the developer has to obtain 100% agreement from the property owners to finalize the land assembly (Shenzhen Municipal Government, 2012). This policy granted infinite power to the affected residents and resulted in power asymmetry between residents and the developer. The power asymmetry caused severe social conflicts and project delays in the developer-led projects, as mentioned in Chapter 5. The local government thus abolished the developer-led mechanism and prevented it applying to other residential redevelopment projects. The development right of the developer has been eliminated. To recapture their control over urban development, through the same policy, the local government reversed the government-led approach in residential redevelopment.

In 2013, the local government promulgated an internal document among the governmental departments to forbid the approval of any residential redevelopment, which withdrew the development right and bargaining power from the affected residents. The market-oriented mechanism was thus technically abolished and only continuously applied to the approved developer-led cases such as Mutoulong. Through these efforts, the local government achieved the reseizure of its intervening power in the urban redevelopment process.

In 2009, the same year when the market-oriented approach was first established, the municipality also issued a government-led redevelopment approach. Considering that market sectors might prioritize the projects with premium location and great commercial value through the market-oriented approach, the local government promulgated the government-led approach as a supplement to redevelop the sites that are ignored by the market. On the basis of the traditional government-monopolized redevelopment mechanism, the government-led mechanism enables the affected residents to negotiate with the local government for fair compensation. In the meantime, the local government maintains its power to initiate a government-led redevelopment project and conduct compulsory acquisition from the last uncooperative residents.

In 2013, in line with the central housing expropriation policy, Shenzhen promulgated its local housing expropriation policy. This policy formally institutionalized the government-led mechanism shaped in 2009. The public participation mechanism significantly empowers the affected residents to negotiate with the local government in the decision-making process. However, the deployment of residents' power was still under the arrangement of the local government because the compulsory acquisition of the state was preserved.

Although the central government has decentralized power to the affected residents and restricted local government intervention through central policies, the local government has adopted diverse local policies to reverse residents' empowerment in order to keep their control over urban redevelopment projects in their own hands. This shows a countermovement in power decentralization. The ability to maintain the desired amount of power in their own hands proves that the local government has increasing political and social autonomy in urban governance. With increasing independence, through local policies, the local government not only materializes the central forces but also enforces its pragmatic demand in the governing process. To achieve its pragmatic demand, the local government sometimes shifts and distorts the central government's requirement at local level. The distortion may result in policy and administrative conflicts between governments in different tiers of the hierarchy.

The local policies fail to effectively and efficiently motivate the market power to facilitate urban redevelopment. The local government primarily focuses on allocating and reallocating the power distribution between its own authority and the residents in order to achieve a delicate power balance for redevelopment processing. The role of the market sectors is highly ignored by the municipality since both developers' power and their involvement are constantly restricted in the decision-making process. However, the expansion of the local governments' autonomy in policymaking increases the risk of policy conflicts at different levels of governments.

Community Attributes

This research argues that both the physical and social attributes of community have great impacts on the structure of power relations. For physical attributes, residents in the communities with premium location, large size and excellent community facilities are granted greater allocative resources when negotiating with the local government or developers. This is consistent with the studies in both marketed-oriented reform (e.g. Hsing, 2010) and participatory redevelopment periods (e.g. Zhang et al., 2016). A new finding is that since the local government draws attention to social stability in participatory redevelopment, the inferior living conditions and poor building quality can add up to the urgency of redevelopment, which grants allocative resources to the affected residents to influence the government's decision on redevelopment.

Social attributes of community that have an impact on power relations include the social status, social bonds, social norms and values of the affected residents. This research agrees with the current studies that residents who work in the local government have more power to negotiate with the local government or the developer (Shin, 2009). On the basis of that, this research further finds that residents with senior ranks in the local government have higher social status. The higher social status grants them greater authoritative resources and allows them to have priority access to relocated housing or higher compensation in redevelopment.

This research agrees with Hsing (2010)'s finding that long-time neighbours nurtured strong social bonds, which could enhance residents' collective actions to counteract against the strategies of the local government and the developer in urban redevelopment. Residents in Ludancun demonstrate the highest level of social bonds in participatory redevelopment among the three cases. The close social relationship substantially contributes to the formation of the self-organized associations. These associations play an important role in leading the affected residents to negotiate with the local government, facilitating the accomplishment of residential redevelopment. In Mutoulong, the social bonds between the affected residents are weaker than that in Ludancun. The construction of self-organized association is seldom. Residents form some temporary coalitions to negotiate with the developer, based on common requirement and long-time neighbours. These coalitions serve to enhance their power to achieve their objectives (e.g. higher compensation, late relocation and double-sided contracts). The social bonds of Nanhuacun are the weakest among the three cases. Because of the huge size of the community and the ambiguous division of property management services, residents in Nanhuacun fail to develop good social relationships with each other. The community mobilization in Nanhuacun is highly fragmented and always in small scale, which triggered the involvement of the real estate agent in participatory redevelopment.

For social attributes of community, current studies ignore the impacts of the social norms and values of the affected residents on power relations. This research finds that social norms are able to add authoritative resources to the affected residents and empower them to take mobilized actions to influence other actors' decisions. For example, the collectivism norm added moral superiority to the cooperative residents. Some cooperative residents formed a coalition and engaged on site to pressurize the uncooperative residents to sign consent.

Values of the affected residents impact their preference towards the redevelopment approach and thus influence their decisions on signing agreements. The participatory redevelopment policies, on one hand, encourage the market sector to play a substantial role in the developer-led redevelopment. On the other hand, through local land use and planning policies, the local government maintains its power to intervene in the government-led redevelopment. The coexistence of these conflictual mechanisms caused unnecessary comparisons between the residents. Since its inception, the developer-led mechanism had provided high compensation schemes to the affected residents, which triggered resistance from the residents in the government-led case. The residents in government-led case adopted waves of mobilization actions to negotiate with the local government to raise their compensation terms. The mobilization ceased until the severe conflicts hindered the proceeding of the developer-led case.

The coexistence of the two mechanisms resulted in a value conflict of the affected resident. Some residents in the developer-led case deem that it is the local government's responsibility to conduct the redevelopment since they bought their apartments from the local government three decades ago. Weak effectiveness of the market-oriented mechanism and the success of government-led case enhances their dependence on the state. Some residents even claim that they will never sign agreements with any private developers. Meanwhile, some residents in the developer-led case regard the residential redevelopment as a certain market decision irrelevant to public interests. Some adopt opportunistic behaviours to maximize their profits by deliberately refusing to sign agreement or requiring unreasonable compensation terms. The conflict of values between these two groups of residents triggered severe conflicts, putting the project at a standstill. Therefore, this research argues that the participatory redevelopment reform resulted in heterogeneous values and preferences among the affected residents. The heterogeneity of values differentiates the decisions of the affected residents and influences the power allocation among different groups.

On the basis of the above discussion, this research finds that the community attributes, composed of physical and social influences, significantly influence the local government's decision on which redevelopment approach is able to apply to one specific participatory redevelopment case. In Ludancun, most of the residents were senior cadres in the local government with great influential power to involve the local government in redevelopment. Based on the strong social bonds, these residents formed the self-organized association as early as in 2003 as a representative to negotiate with the local government. Residents in Ludancun possess the same value that the state is more reliable and trustworthy than market sectors so that they adopted strategies to turn down the engagement of private sectors. The social status, social bonds and common value contribute to the application of the government-led approach in Ludancun redevelopment.

After the release of the draft market-oriented redevelopment policy, being attracted by the premium location, large size and excellent facilities of the community, Yitian began to engage in Mutoulong redevelopment. The developer provided the highest ever compensation terms and acquired ample supports from the majority of the affected residents in a short period. Consequently, Yitian successfully applied for the developer-led agenda and became the redevelopment candidate. However, the discrepancy of residents' values and preferences in Mutoulong resulted in severe conflicts, resulting in project delay. In this regard, physical attributes, including the premium location and large size of the community, significantly contribute to the application of the developer-led approach in Mutoulong.

In Nanhuacun, after the establishment of the market-oriented redevelopment policy, the premium location and large size of the community attracted multiple developers engaging on site to collect redevelopment support. However, the dearth of social bonds and the multi-developer involvement in the community differentiated residents' preferences. With no developer obtaining enough supports from the residents, Nanhuacun, therefore, failed to be listed in the developer-led redevelopment agenda. Since the developer-led approach was abolished, the owner-led redevelopment approach became the only approach to conduct residential redevelopment. Considering that the owner-led redevelopment once occurred in Nanhuacun, the real estate agents engaged on site and guided the residents to apply to the local government for the self-redevelopment agenda. The physical and social attributes of the community jointly influenced the applied redevelopment approach in Nanhuacun.

7.1.3 Discussion on Endogenous Influences

Local Government

The power that local government possesses originates from structural resources (ruling position, land ownership, definition of regulations and policies and jurisdictions on land use and development intensity), allocative resources (capital investment) and authoritative resources (social relationships) in participatory redevelopment, which is consistent with the one in the marketed-oriented reform period. However, compared to its peers in the marketed-oriented reform period, the local government scarcely obtains its power resources from the control over financial institutions in this research.

Because of their ruling positions, local governments resort to the regulatory authority over redevelopment planning to approve, suspend or cancel the project (by applying choice rule to constrain other actors' actions), to set up requirement for the developer in the land auction process (by applying boundary rule to define how developer is chosen) and to determine which developer can join in the decision-making process (by applying position rule to specify position for developer). The jurisdictions on land use and development intensity enable it to adjust the redevelopment plan to influence the interest distribution among the main stakeholders (by applying pay-off rule to distribute interests among participants). Facing the last uncooperative households, the local government can compulsorily expropriate their housing and finalize the property assembly in the government-led case (by applying scope rule to forbid the resistance of the uncooperative households). Through setting the redevelopment project as a key project, the local government is also able to mobilize more financial resources and share more interests with the affected residents to facilitate its implementation (by applying pay-off rule to distribute interests between the government and residents). These strategies are consistent with those adopted by the local government in the marketed-oriented reform period. All these rules serve to take effect in practice, based on the dominant power the local government possessed.

Nevertheless, there are three new shifts about the rules of the local government in participatory redevelopment. Firstly, the local government often utilizes public media to publicize the detailed redevelopment plan and compensation terms to collect comments from the affected residents (by applying information rule to define public media as a new communication approach). This rarely happened in the socialist and the marketed-oriented reform period. This is because the participatory redevelopment reform establishes the public participation mechanism and increases the requirement of process transparency in urban redevelopment.

Secondly, to facilitate the redevelopment, the local state establishes a local public agent as a state-endorsed representative on site (this is an application of the position rule), as it often did in the marketed-oriented reform period. Nevertheless, the agent is different from the government-owned development company or other entrepreneurial institutes which exclusively aimed to seek financial gains from the project in the marketed-oriented reform period. In this research, the agent works as a local communicating channel to facilitate the negotiation between the local government and the affected residents without pressure for profits. In the marketed-oriented reform period, the local government heavily relied on the monetary revenue from the land lease, so the local agent aimed to accelerate the project for profit. In participatory redevelopment, the land-related revenue for the local government is significantly decreased. Instead, the government requires free urbanized land, public housing and infrastructures from the urban redevelopment projects. As the measure that the local government benefits from participatory redevelopment has been changed, the government decreases its profit demand in local cases.

Thirdly, the local government in participatory redevelopment seldomly resorts to the aggregation rule to form coalitions with other stakeholders. Local governments in marketed-oriented reform redevelopment used to form an alliance with other resource-rich actors such as the international developers to address the shortage of expertise and financial resources (e.g. Yang & Chang, 2007). In the urban village redevelopment, the local government formed a coalition with the de facto land owner—the village collective to facilitate the redevelopment. In this research, neither the developer nor the residents are able to obtain, maintain or strengthen their power by forming a coalition with the local government. The difference manifests that the local government becomes increasingly independent and powerful in participatory redevelopment practice.

This research attributes the independence of the local government for three reasons. Firstly, the local government in Shenzhen has become rich in financial resources. The marketed-oriented reform in the late 1980s has nurtured a vitalised and well-developed market economy in Shenzhen. The local government acquired ample tax-related revenue from the market sector and thus becomes a stakeholder with sufficient financial resources. Coupled with its strong structural resources, the local government per se has been powerful enough to dominate the local urban development and no longer has to form coalitions with other financially powerful stakeholders to enhance its domination. Secondly, with the boom of property development in China, more and more domestic developers grow into professional developers with ample expertise and financial capitals. The local government does not have to form an alliance with international developers to complete the redevelopment. Thirdly, as the landowner of the dilapidated residential area, the local government has great control over residential redevelopment. So, the government has no intention to form a coalition with the affected residents to seek the latter's cooperation. This is different from the urban village where the village collective possesses land ownership.

Residents

Consistent with residents in the marketed-oriented reform period, the affected residents of participatory redevelopment are still in possession of power resources from working in local government and other public institutes, involved in the hallmark projects and long-time neighbours. However, compared with the residents in the marketed-oriented reform period, residents in participatory redevelopment are in possession of increasing power resources. The institutionalized property rights grant strong structural resources for the residents and empower them not only to oppose compensation terms they deem unfair but also to profit from the private property ownership. The public participation mechanism protects residents' right to participate in the decision-making process to negotiate for higher compensation (Li & Liu, 2018). Moreover, authoritative resources, such as reputation and higher social status, significantly empower them to adopt diverse strategies and actions to influence the decision-making process. For example, some residents in Nanhuacun are professors or senior engineers with strong personal influences and reputation. They are able to wield their authoritative resources to affect the decision of other residents and pressurize the local government for prompt response to local petition. Ludancun residents who were senior cadres in the local government have maintained a close relationship with governmental cadres. They are able to influence the decision-making of the local government.

Villagers exhibit place attachment in urban village redevelopment because ancestral houses, hall and even graves are located on site (Chung, 2013). The place attachment added to their allocative resources when negotiating with the local government and developers (Chung, 2013). In this research, residents live in dilapidated residential areas manifest their place attachment through the reliance on the surrounding healthcare and education facilities. The place attachment also empowered them to negotiate with the local government and developers as allocative resources.

Compared to villagers in urban village redevelopment, residents in the dilapidated residential area were not authorized with the self-redevelopment rights as structural resources. Although self-redevelopment rights were authorized by the central policies, these rights are invalid to the affected residents in Shenzhen. This research attributes residents' absence of self-redevelopment rights to their lack of land ownership. In urban village, as the land is collectively owned by the village, the local government has to make great concession to the villagers by authorizing self-redevelopment power to smooth the way of urban redevelopment. In this regard, self-redevelopment signifies greater extent of autonomy and larger share of land value increment to the villagers. On completion of the redevelopment project, the collective-owned land transforms to be state-owned land and the local government also significantly benefits from the process with more urbanized land resources under its control (Hsing, 2010). However, residential redevelopment manifests a completely different scenario. Since the land in residential area is state-owned, the government possesses significant authority on land disposal, so they have no desire to authorize self-redevelopment rights to the affected residents to promote redevelopment. In addition, the profit to the local government in dilapidated residential redevelopment is far lesser compared to its benefit from urban village redevelopment, which also diminishes its intention for owner-led residential redevelopment.

Residents also resort to some strategies and actions to exert and enhance their power in participatory redevelopment. Consistent with the peers in the marketed-oriented reform period (Hsing, 2010), residents in participatory redevelopment also conduct self-taught of relevant policies. This helps them have a better understanding of how to communicate with the local government and developers (by applying information rule) to maximise their own interests within a reasonable legalized framework in participatory redevelopment. In the meantime, they also attempt to pressurize the local government and developers through lodging individual and collective actions, such as protests, petitions and the invitation of public media (by applying choice rules to constrain the local government and developers actions), as the residents in the marketed-oriented reform and participatory period (Cheng, 2012; Hsing, 2010). Social media has become a critical platform for the residents to conduct social mobilization and make a claim on their rights (by applying information rule to set up a new publicity approach).

To strengthen their bargaining power, residents form coalitions with other residents or agents to negotiate with the local government or the developer in participatory redevelopment (this is the application of aggregation rule). In Mutoulong, 30 tough residents established a coalition to bargain with the developer, based on long-time neighbourhood and same interest demands. In Nanhuacun, residents formed a coalition with the real estate agent to apply to the local government for the self-redevelopment agenda.

With newly configurated power resources, residents commence taking some new strategic actions which are seldomly observed in the marketed-oriented reform period. The property right reform enables residents to perceive their properties as commodities for commercial benefits while the participatory redevelopment reform constrains the power of the local government and authorizes bargaining power to residents. Some residents begin to take opportunistic behaviours, such as signing double-sided contracts add deliberately rejecting the developer's compensation offer, to maximise their economic returns in participatory redevelopment (by applying pay-off rule to redistribute the interests). These behaviours severely distort the negotiation process and result in project delay, which is seldomly observed in the existing research. This finding agrees with Schoon (2014)'s argument that the loss of regulatory power in China's urban redevelopment increased informal activities and increased temporary systemic instability.

Compared with current studies, although residents submitted collective petitions to governments at higher level for political assistance, the direct intervention of the higher-level government is rare in Shenzhen. Current studies prove that the involvement of the central government normally occurs in the projects which have a significant impact on social stability. For example, the urban regeneration of the Drum Tower Muslim District is highly related to sensitive social issues about religion and ethnicity (Zhai & Ng, 2013). In the urban redevelopment of Jiuxiaoqian in Beijing, the central government directly intervened to smooth the way of the 2008 Olympic Games. In comparison, dilapidated residential redevelopment in Shenzhen does not display any linkages to sensitive social issues, nor negatively impact international mega-events. Although local residents keep petitioning to governments at higher level in Mutoulong and Nanhuacun, high-level governments feel reluctant to get involved. The mounting power of the affected residents enabled them to obtain better compensation from urban redevelopment projects in the participatory period.

In the marketed-oriented reform period, residents were either required by the local government to pay for on-site relocation (Shin, 2009), or arranged to relocate to in-situ housing in periphery areas (Yang & Chang, 2007), or paid with monetary reimbursement which was far below the price of newly constructed on-site apartments (Yang & Chang, 2007). In this research, the residential redevelopment may still temporarily negatively impact their job and facility accessibility as the affected residents complained during the demolition and construction period. However, residential displacement no longer increases the economic burden on the original residents. This is because the residents are not required to pay for redevelopment but can obtain subsidy covering rental housing, interior decoration loss and relocation. In addition, it will not adversely affect housing affordability of the original households because all the households with legal property ownership are able to acquire on-site relocated apartments with larger size. In this regard, this research finds that the participatory reform indeed plays an important role in protecting private property rights and interests of the affected residents in the redevelopment process.

Developer

In this research, developers primarily obtain their power from allocative resources (expertise, financial assets, reputation and experiences) and authoritative resources (a close relationship with the local government), the same with developers in the marketed-oriented reform period. Compared to the peers in the marketed-oriented reform period, developers in participatory redevelopment are granted the development right by the participatory policy as a new structural resource. In addition, supports from the vast majority of the affected residents can work as authoritative resources to the developer when it negotiates with the opposing residents.

Developers take some strategies and actions to achieve their objectives by influencing others' actions. In the marketed-oriented reform period, developers always hire demolition companies to adopt forceful strategies to compel the residents to relocation (by applying choice rule to constrain residents actions), which accelerated the demolition and relocation process (Wu, 2004). The developer formed a coalition with the local government to facilitate the enforced relocation process, based on the good relationship with the latter. However, in this research, a close relationship with some government cadres indeed empowered the developer to take opportunistic behaviours to pressurize the affected residents (by applying choice rule to constrain residents' actions) in developer-led redevelopment. The developer also adopted forceful strategies to pressurize the opposing residents (by applying choice rule to constrain residents' actions). However, without the endorsement of the local government, the developer has no power for the compulsory acquisition. The forceful behaviours thus cannot smooth the way for redevelopment as they used to in the marketed-oriented reform period. In contrast, these behaviours triggered severe conflicts between the developer and the residents, putting the project at a standstill.

There are fewer categories of developers involved in participatory redevelopment, compared to the marketed-oriented reform period. All developers in the three cases are Chinese domestic private developers, while the engagement of the government-owned developers or international developers is rare. The boom of the real estate market in China nurtures powerful domestic private developers. In this research, some of them were previously government-owned developers and became private companies through a privatized process, such as the UCDC. Some were originally small local private developers, such as EG and Vanke. These companies have grown into large developers with ample expertise, capital resources and advance construction techniques. Because of local knowledge, social relationships and development experience, they are able to exclude the involvement of international developers in participatory redevelopment.

Real Estate Agent

As a newly emerged actor in participatory redevelopment, the real estate agent possesses few structural power but primarily procures its power from financial strength as allocative resources and reputation and close social relationship with the affected residents as authoritative resources. It possesses no structural resource since its existence and operation are in the "grey legal zone". The agent makes efforts to escalate compensation terms (by applying pay-off rule) and develop a close relationship with the residents (by applying aggregation rule to unite the residents) to assemble residents' supports. It organized a property owner committee to mobilize the affected residents (by applying aggregation rule to unite the residents) to enhance its power. Meanwhile, it conducts building quality test to prove the poor living condition and invites the media to publish the test result (by applying information rule), in order to pressurize the government (by applying choice rule to constrain government actions). However, all of these rules fail to deploy the residents' self-redevelopment power for owner-led redevelopment.

7.1.4 Connecting Power, Power Relations and Institutions

The above discussion on exogenous and endogenous influences contributes to connecting power, power relations and institutions in this research. Power exercised by one actor, in terms of rules, can influence other actors in various ways, producing different effects depending on the context and situation. The deployment of power relies on the configuration of power resources, which function as mediums enabling actors to act within the decision-making process. In other words, resources discharge acting capacities to actors, allowing them to consciously pursue their own objectives. This study aligns with Foucault's theory of power, which posits that power configurations emerge only within social interrelationships when multiple actors engage in mutual interactions. The exertion of power by one actor inevitably impacts counterparts and generates differences in outcomes.

Among the three types of resources examined, structural resources predominantly determine an actor's position and, consequently, the overall power structure among stakeholders. The local government occupies the highest position due to its superior structural resources. This privileged position enables it to deploy various rules to dominate the participatory redevelopment process. Affected residents with strong structural resources hold a secondary position, enabling them to adopt diverse strategies to negotiate with resource-rich actors for fair compensation. Developers, possessing relatively few structural resources, occupy a tertiary position, with limited strategies to influence decision-making. Real estate agents, lacking structural resources, occupy the lowest position and are unable to significantly affect the decision-making process despite deploying multiple strategies.

Exogenous influences, particularly institutional arrangements, largely determine the distribution of structural resources among actors. Structural properties—central policies, local land use and planning regulations and community attributes—exert top-down forces that both enable and constrain actors' behaviours. Central policies gradually reduce the structural resources of local governments while granting substantial resources to affected residents, enabling their participation in decision-making through mechanisms such as institutionalized property rights and public participation channels. However, the local government both materializes central directives through market-oriented redevelopment mechanisms and preserves its intervention via government-led approaches. When market-oriented mechanisms perform poorly, the government may abruptly abolish them and revert to government-led strategies, strategically limiting residents' empowerment while maintaining control over redevelopment projects. Thus, central and local participatory redevelopment policies collectively shape the mechanisms applied in Shenzhen, which in turn frame the power structure among stakeholders. Community attributes further determine which redevelopment mechanism applies in specific cases. In short, institutional forces strongly shape the power structure in participatory redevelopment, rendering power relations hierarchical due to their dependence on structural properties.

Although institutional arrangements shape the power structure through resource configurations and rules, actors retain degrees of autonomy and the capacity to resist.

Interactions among actors constitute a bottom-up force that can modify institutions, revealing adaptation and reproduction in response to structural constraints. For example, in government-led cases, affected residents, guided by the Property Owners Committee, conducted multiple negotiation rounds in the equal consultation arena, successfully securing higher compensation. However, not all resistance efforts were effective. In developer-led cases, developers with authoritative support from most residents, as well as abundant financial and technical resources, attempted to coerce uncooperative residents. These strategies failed to succeed, halting the project. Similarly, in owner-led cases, real estate agents mobilized residents to request redevelopment agendas from local government, yet the government countered these efforts. These cases demonstrate that the accumulation of allocative or authoritative resources by lower-position actors does not alter their structural position or ensure strategy effectiveness. Thus, an actor's position is determined by the character of its structural resources rather than their quantity.

The success of residents in government-led cases can be explained by three factors. First, many affected residents were senior government cadres, and the Ludancun redevelopment still allocated apartments according to cadre rank, reflecting path dependence from the socialist welfare system. Higher-ranked cadres could secure larger apartments. Second, some residents held esteemed reputations and close relationships with current officials, allowing them to influence decision-making. Third, residents consistently maintained a cooperative attitude during negotiations, avoiding conflict and establishing a stable, collaborative relationship with the government. These findings highlight that being a senior cadre in China constitutes a powerful authoritative resource and that stable, collaborative relationships facilitate successful negotiation outcomes (Zhang et al., 2016).

The effectiveness of rules is largely determined by an actor's position. Rules imposed from higher- to lower-position actors generally take effect, particularly when interests align, facilitating smooth implementation. For instance, when the government increases compensation to secure residents' agreement in government-led redevelopment, the pay-off rule operates effectively. When interests conflict, rule enforcement entails higher costs, as reflected in the government's efforts to counteract mobilization by residents and real estate agents in owner-led cases. Conversely, rules from lower- to higher-position actors take effect only when they align with the higher-position actor's interests. For example, in developer-led cases, sincere negotiation strategies addressing residents' genuine demands allowed the information rule to secure agreements. High-frequency rule application may reinforce the rule itself but does not guarantee successful outcomes, as demonstrated by the failure of developers' repeated use of choice rules to exclude opposing residents.

Through the connection between power, power relations and institutions, this research finds that institutional arrangements primarily determine the distribution and redistribution of structural resources among stakeholders, as exogenous influences. Structural resources define an actor's position, while accumulations of allocative or authoritative resources cannot alter it. An actor's position further dictates the effectiveness of its rules: rules from higher- to lower-position actors generally take effect, but conflicting interests raise enforcement costs, whereas rules from lower- to higher-position actors succeed only if they serve the higher actor's objectives. High rule frequency strengthens the rule itself but does not guarantee action outcomes.

Across different cases indicates that urban redevelopment in China is not solely elitist, contrary to Hunter (1953) and aligns with Dahl's (1961) pluralism, where multiple stakeholders can influence decision-making. Even within an authoritarian regime, the government cannot fully control the process without distributing power to actors such as developers and residents. Consequently, the government frames redevelopment agendas as common public objectives, claiming broad benefits for all actors. Nonetheless, groups engage differently in decision-making and achieve varying levels of gain. This study supports Foucault's (1982) argument that power relations arise from interactions where actors exercise power over others who are neither inert nor powerless.

Finally, structural configurations and agent interactions reciprocally shape power relations, consistent with Giddens's (1984) duality of structure. This research finds that in China, structural properties exert a stronger influence than agents' interactions in shaping power relations. Exogenous influences, as structural properties, form institutional arrangements that determine actor hierarchies. Central and local policies define the distribution of structural resources, while community attributes determine the redevelopment approach applied in projects. Actors mobilize various power resources and interact with one another, producing and reproducing power relations and influencing participatory redevelopment outcomes. However, these interactions rarely overturn the power structure, indicating that power relations in participatory redevelopment are largely determined and constrained by the policy framework, which is dominantly controlled by the state. The findings are consistent with Perry (2008)s argument that

> In a country (China) where rights are seen more as state-authorized channels to enhance national unity and prosperity than as naturally endowed protections against the state intrusion, popular demands for the exercise of political rights are perhaps better seen as an affirmation of - rather than an affront to - state power (pp. 46–47)

7.2 Summary

By connecting the empirical findings of this research to the existing literature, this chapter provides an in-depth discussion of power relations, as well as exogenous and endogenous influences. Four main stakeholders are identified: the local government, affected residents, developers, and real estate agents. Rather than serving a merely facilitative role, the local government continues to play a dominant role, exercising the strongest control over participatory redevelopment. Residents are increasingly influential in decision-making, with growing bargaining power; however, their empowerment remains heavily contingent upon the local government's approval. Without governmental endorsement, developers' influence is constrained, jeopardizing their financial feasibility in urban redevelopment. Concurrently, a new market actor—the real estate agent—emerges in participatory redevelopment, attempting to act as a power broker for residents to leverage their "self-redevelopment power". Since this self-redevelopment power proves largely ineffective for residents, real estate agents remain marginalized in the decision-making process despite employing various strategies and actions. Consequently, overall power relations in participatory redevelopment remain conflictual and fragmented, rather than inclusive or collaborative. The next chapter will summarize the main findings and present the dissertation's conclusions.

References

Abramson, D. (2011). Transitional property rights and local developmental history in China. *Urban Studies, 48*(3), 553–568.

Chen, Y., Zhang, X., Chau, K. W., & Yang, L. (2022). How the institutional change in urban redevelopment affects the duration of land redevelopment approval in China? *Land Use Policy, 119*, 106160. https://doi.org/10.1016/j.landusepol.2022.106160

Cheng, Z. (2012). The changing and different patterns of urban redevelopment in China: A study of three inner-city neighborhoods. *Community Development, 43*(4), 430–450. https://doi.org/10.1080/15575330.2012.711763

Chung, H. (2013). The spatial dimension of negotiated power relations and social justice in the redevelopment of villages-in-the-city in China. *Environment and Planning A, 45*(10), 2459–2476. https://doi.org/10.1068/a45416

Dahl, R. A. (1961). *Who governs?: Democracy and power in an American city*. Yale University Press.

Foucault, M. (1982). The subject and power. *Critical Inquiry, 8*(4), 777–795.

Giddens, A. (1984). *The constitution of society: Outline of the theory of structuration*. Univ of California Press.

Han, S. S., & Wang, Y. (2003). The institutional structure of a property market in inland China: Chongqing. *Urban Studies, 40*(1), 91–112.

He, S., & Wu, F. (2005). Property-led redevelopment in post-reform China: A case study of Xintiandi redevelopment project in Shanghai. *Journal of Urban Affairs, 27*(1), 1–23. https://doi.org/10.1111/j.0735-2166.2005.00222.x

He, S., & Wu, F. (2009). China's emerging Neoliberal Urbanism: Perspectives from urban redevelopment. *Antipode, 41*(2), 282–304. https://doi.org/10.1111/j.1467-8330.2009.00673.x

Hsing, Y.-T. (2010). The great urban transformation: Politics of land and property in China. *Oxford University Press*. https://doi.org/10.1093/acprof:oso/9780199568048.001.0001

Hunter, F. (1953). *Community power structure A study of decision makers*. The University of North Carolina Press.

Li, B., & Liu, C. (2018). Emerging selective regimes in a fragmented authoritarian environment: The 'three old redevelopment'policy in Guangzhou, China from 2009 to 2014. *Urban Studies*, 0042098017716846.

Li, B., Tong, D., Wu, Y., & Li, G. (2019). Government-backed 'laundering of the grey'in upgrading urban village properties: Ningmeng Apartment Project in Shuiwei Village, Shenzhen, China. *Progress in Planning*, 100436.

Li, L., & Li, X. (2011). Redevelopment of urban villages in Shenzhen, China—An analysis of power relations and urban coalitions. *Habitat International, 35*(3), 426–434. https://doi.org/10.1016/j.habitatint.2010.12.001

Mertha, A. (2009). "Fragmented authoritarianism 2.0": Political pluralization in the Chinese policy process. *The China Quarterly, 200*, 995–1012.

Perry, E. J. (2008). Chinese conceptions of "rights": From Mencius to Mao—and now. *Perspectives on Politics, 6*(1), 37–50.

Po, L. (2011). Property rights reforms and changing grassroots governance in China's urban—Rural peripheries: The case of changping district in Beijing. *Urban Studies, 48*(3), 509–528.

Schoon, S. (2014). Chinese strategies of experimental governance. The underlying forces influencing urban restructuring in the Pearl River Delta. *Cities, 41*, 194–199. https://doi.org/10.1016/j.cities.2014.01.008

Shenzhen Municipal Government. (2012). *Detailed Rules for Urban Redevelopment implementation in Shenzhen.*

Shin, H. B. (2009). Residential redevelopment and the entrepreneurial local state: The implications of Beijing's shifting emphasis on urban redevelopment policies. *Urban Studies, 46*(13), 2815–2839.

Tian, L., & Yao, Z. (2018). From state-dominant to bottom-up redevelopment: Can institutional change facilitate urban and rural redevelopment in China. *Cities, 76*, 72–83.

Wu, F. (2004). Residential relocation under market-oriented redevelopment: The process and outcomes in urban China. *Geoforum, 35*(4), 453–470.

Wu, F. (2018). Planning centrality, market instruments: Governing Chinese urban transformation under state entrepreneurialism. *Urban Studies*, 0042098017721828.

Wu, F., Zhang, F., & Liu, Y. (2022). Beyond growth machine politics: Understanding state politics and national political mandates in China's Urban Redevelopment. *Antipode, 54*(2), 608–628. https://doi.org/10.1111/anti.12781

Yang, Y.-R., & Chang, C.-H. (2007). An urban regeneration regime in China: A case study of urban redevelopment in Shanghai's Taipingqiao area. *Urban Studies, 44*(9), 1809–1826. https://doi.org/10.1080/00420980701507787

Ye, L. (2011). Urban regeneration in China: Policy, development, and issues. *Local Economy,26*(5), 337–347. https://doi.org/10.1177/0269094211409117

Yeh, A. G., & Wu, F. L. (1996). The new land development process and urban development in Chinese cities. *International Journal of Urban and Regional Research, 20*(2), 330–340.

Zhai, B., & Ng, M. K. (2013). Urban regeneration and social capital in China: A case study of the Drum Tower Muslim District in Xi'an. *Cities, 35*, 14–25.

Zhang, L., Chen, J., & Tochen, R. M. (2016). Shifts in governance modes in urban redevelopment: A case study of Beijing's Jiuxianqiao Area. *Cities, 53*, 61–69. https://doi.org/10.1016/j.cities.2016.01.001

Zhang, T. (2002). Urban development and a socialist pro-growth coalition in Shanghai. *Urban Affairs Review, 37*(4), 475–499. https://doi.org/10.1177/10780870222185432

Zhou, Z. (2014). Towards collaborative approach? Investigating the regeneration of urban village in Guangzhou, China. *Habitat International, 44*, 297–305.

Zhu, J. (1999). Local growth coalition: The context and implications of China's gradualist urban land reforms. *International Journal of Urban and Regional Research, 23*(3), 534–548.

Chapter 8
Conclusion

8.1 Summary of Findings

This research has identified the key stakeholders, examined their power relations, and assessed the exogenous and endogenous influences that drive changes in these relations within participatory redevelopment. The following section summarizes the major findings of the book.

a. *Residents have been substantially empowered by the participatory redevelopment policies to actively join in the decision-making process.*

1. Residents can participate in the decision-making process and influence local government's and developers' decisions to bargain for fair compensation in urban redevelopment. This finding accords with current studies that residents begin to play an essential role in the decision-making process in participatory redevelopment (e.g. Zhang et al., 2016);
2. However, this study shows that residents' empowerment is highly fragmented and differentiated. Households whose properties have been enlisted for redevelopment are substantially empowered to bargain for fair compensation. Households whose properties are not scheduled for redevelopment are excluded from both getting more power and participating in the decision-making process;
3. Current literature fails to elaborate on the reasons for residents' empowerment. This research finds that residents are significantly empowered by two mechanisms in participatory redevelopment policies. The institutionalized property rights empower them not only to oppose compensation terms they deem unfair but also to profit from their private property ownership. The public participation mechanism protects their rights to participate in the decision-making process to negotiate for higher compensation;
4. Policy conflicts between governments in different tiers of the hierarchy may significantly empower the affected residents in participatory redevelopment. Residents thus adopted speculative behaviours to maximise their profits, which

X. Li, *Participatory Redevelopment as Contested Arenas in China*,
https://doi.org/10.1007/978-981-95-9394-1_8

resulted in severe project delays. These findings have not previously recorded in the literature;

5. A new finding reveals that the affected residents fail to be authorized the self-redevelopment right to start owner-led redevelopment in participatory redevelopment. This research provides an explanation that the collective and/or state rather than private land-ownership arrangement leads to the absence of residents' self-redevelopment right;
6. This research agrees with the literature that residents always exert individual resources to enhance their influences in the decision-making process, such as reputation, social status and social bonds (e.g. Hsing, 2010). Besides, this research further finds that based on the power granted by participatory redevelopment policies, these resources are more effective than before in empowering residents to achieve their aims in urban redevelopment;
7. Consistent with the literature, this research finds residents' increased power enables them to obtain satisfying compensation terms in urban redevelopment. However, this research finds that Shenzhen's participatory redevelopment policies grant higher power to the affected residents to perform better in property rights protection. Residents are not required to pay for reconstruction but can obtain on-site newly constructed apartments with larger size and monetary reimbursement covering the cost of rental housing, interior decoration and relocation;
8. Since the affected residents are empowered in participatory redevelopment, it becomes necessary for both the local state and developers to obtain supports from the majority of the affected residents to complete land assembly. This finding matches that observed in earlier studies (e.g. Zhang et al., 2016);
9. Compared with normal ones, residents who are enrolled in the local government have greater power to influence the local government's decision and are more likely to obtain higher compensation terms in urban redevelopment, which corroborates the current studies (e.g. Hsing, 2010);
10. A new finding in this research suggests that residents with senior ranks in the government have prior access to relocated housing and higher compensation terms in urban redevelopment. This is regarded as the path dependence of the socialist regime;
11. The place attachment of the affected residents in dilapidated residential area manifests in their reliance on surrounding education and health care facilities. It is rather different from villagers in the urban village, where the ancestral houses, hall and graves are located on site (Chung, 2013);
12. This research confirms that residents always resort to media power to strengthen their influence when negotiating with the local government or developers. On the basis of that, this research further finds that with the development of the internet, social media has become a critical platform for the residents to initiate mobilization and make a claim on their rights in urban redevelopment;
13. Earlier research finds that because of residents' petitions, the central government directly engage in influencing the local government's decisions in local redevelopment projects (e.g. Zhai & Ng, 2013; Zhang et al., 2016). In contrast to earlier

studies, however, there is not any direct intervention of higher-level government in dilapidated residential redevelopment in Shenzhen, despite continuous petitions from the residents. The absence of intervention in higher-level governments is because the residential redevelopment projects do not display any linkages to sensitive social issues, nor negatively impact international mega-events;

b. *Local government assumes the most powerful role and becomes more independent in participatory redevelopment.*

14. The local government's overwhelming power has been preceded by the mounting community power. Therefore, the local government fails to implement the redevelopment process according to its own determination by forceful strategies and actions. This is consistent with current studies (e.g. Chung, 2013; Zhou, 2014);
15. In contrast to many current studies (e.g., Chung, 2013; Zhou, 2014), however, this research finds that the local government still plays the most powerful role with the strongest control over participatory redevelopment. This supports the persisting domination of state power in China's urban development as argued in Han (2000);
16. The strict constraints over rural land acquisition increase the importance of land resources in participatory redevelopment. The land ownership becomes a powerful resource which significantly empowers the local government to control the redevelopment project in all the major perspectives, including the decision of redevelopment approach, the approval of redevelopment agenda and redevelopment plan. These decisions significantly influence the power structure and interest distribution among the involved stakeholders. These findings are inconsistent with the argument that the local government plays a facilitating and conducive role in participatory redevelopment;
17. A new finding in this research reveals that with increasing autonomy in making local policies, the local government utilizes local policies to enforce its pragmatic demands and maintain its intervention in urban development. These policies sometimes shift and distort the central government's requirement;
18. The local government, through the strategic deployment of its resources and rules in the interaction process, cunningly counteracts other stakeholders' power and enhances its dominant power in participatory redevelopment. This finding has not previously been depicted in participatory redevelopment;
19. Current studies revealed that the local government had to form a coalition with other resources-rich actors to facilitate urban redevelopment, such as international developers with development experience, expertise and financial resources and bank institutions with financial resources (e.g. Han & Wang, 2003; Zhang, 2002). However, in contrast to earlier findings, this research finds that the local government has no intention to form an alliance with these actors and becomes more independent in participatory redevelopment;
20. Relevant literature contends that participatory redevelopment policies promote a fair and transparent decision-making process (e.g. Li & Liu, 2018). The local government thus commences utilizing public media to publicize the detailed

redevelopment plan and compensation terms to collect comments from the affected residents. This is rarely discussed in the literature;

21. The new land transfer mechanism changes how local governments benefit from urban redevelopment. The local government turns to require free urbanized land, public housing and infrastructures from participatory redevelopment projects rather than land-based monetary revenues. Therefore, the local state decreases its profit demand and commences paying more attention to residents' welfare and social stability in local projects. These findings have not previously been discovered;

c. *Developers' power is constrained by the local government's strong power and the mounting community power.*

22. There are evident knowledge gaps about the role and power of developers in participatory redevelopment. This research finds that the role of developers becomes more differentiated in participatory redevelopment. The private developer has been authorized redevelopment right by the participatory redevelopment policies. They can initiate and carry out redevelopment by directly negotiating with the affected residents in the developer-led projects. In the government-led projects, the developer's involvement is postponed playing a builder's role with little power to influence the decision-making process;
23. Developers' power becomes much weaker in participatory redevelopment and their power is constrained by the powerful local government and the residents with increased power. The decreased power negatively impacts the financial feasibility of the developers. These findings are seldomly revealed in the literature;
24. Consistent with current studies (e.g. Wu, 2004), this research finds that developers not only adopt forceful strategies but also hire the demolition company to pressurize the uncooperative residents to be cooperative. However, this research finds that without the local government's endorsement, the developer lacks the power to counteract the uncooperative residents' power, despite these enforced strategies;
25. A new finding suggests that international developers seldom engage in participatory redevelopment because domestic developers have grown to be financially powerful and expertise-rich in China;

d. *Real estate agent, as a new actor, emerges to act as a power broker of the affected residents.*

26. This research identifies a new market player—the real estate agent—in the decision-making process. The agent attempts to play a power broker role to deploy residents' self-redevelopment power in participatory redevelopment;
27. A new finding suggests that the absence of a self-organized community association (e.g. property owner committee) organizing community affairs results in the engagement of the real estate agent;
28. The real estate agent fails to receive power from the participatory redevelopment policies, so it has no structural power. It mainly procures its power from financial resources, reputation and close social relationships with residents. Meanwhile, the agent mobilizes the affected residents and resorts to media power to

enhance its power in participatory redevelopment. However, all these strategies fail to come to force. These findings have not previously been explained in the literature;

29. The emergence of the real estate agent attests the empowerment of the affected residents. This is consistent with the literature that agents only come to work as a local representative for the powerful actors in the decision-making process;

e. *The growing number of stakeholders adds to the complexity of participatory redevelopment.*

30. The empowerment of the affected residents increases the number of stakeholders in participatory redevelopment and adds to the complexity of property rights and interests in the decision-making process. Different from current studies (e.g. Zhang et al., 2016; Zhou, 2014), power relations in participatory redevelopment become more chaotic and conflictual, rather than collaborative and inclusive;
31. A new finding reveals that participatory redevelopment has become a highly specialized task to both developers and the local government, requiring excellent communication skills, enormous manpower, capital investment and effective risk-tackling strategies;
32. Regarding the dilapidated residential redevelopment, government-led approaches are considered more effective than market-led approaches in addressing the fragmentation and complexity of property rights. This is primarily because government intervention can effectively manage property owners' heterogeneity, thereby reducing project delays (Li et al., 2024). However, this approach typically demands a substantial allocation of administrative resources, rendering it difficult to scale up.

f. *The central government and the local government exhibit divergent priorities in the reform of participatory redevelopment, reflected in conflicting policy directions.*

33. To pacify social resistance and facilitate urban redevelopment, the central policies gradually constrain the local government's intervention and encourage the local government to play a conductive and regulatory role in participatory redevelopment. Meanwhile, these policies grant development right to the developer and self-redevelopment right to the land/property owners, launching a power decentralization process. These findings are in agreement with current studies (e.g. Li & Liu, 2018; Tian & Yao, 2018);
34. The influences of local policies seldomly examined in the literature. This research finds that through local land use and planning policies, the local government strives to maintain its control over urban redevelopment projects and constrain the empowerment of the affected residents, showing a countermovement of the power decentralization;
35. The expansion of the local governments' autonomy in policymaking increases the risk of policy conflicts between different levels of governments. These findings have not previously been described in the literature;

g. *Community attributes play a significant role in determining the choice of redevelopment approach for a given project.*

36. This research agrees with the current research that communities with premium location, large size and excellent facilities can empower the affected residents when they negotiate with the local government or the developer (Hsing, 2010). Since the local government draws attention to social stability in participatory redevelopment, this research further finds that the inferior living conditions and poor building quality can add to the urgency of redevelopment, which influences the government's decision on redevelopment;
37. This research firstly finds that residents in the community with better social bonds are more likely to form self-organized associations. These associations play an essential role to lead the collective mobilization, which facilitates the success of urban redevelopment;
38. A new finding reveals that social norms are able to add authoritative resources to the affected residents and empower them to take mobilized actions to influence other actors' decisions;
39. This research finds that values of the affected residents impact their preference towards the redevelopment approach and thus influence their decisions on signing agreements, which has not been discovered in the literature;

h. *Power relations in participatory redevelopment are hierarchical, mainly shaped by the institutional arrangement.*

40. Power is never dominated by a single, coherent elite even in an authoritarian environment in China. However, it neither equally distributes across all social groups;
41. The power structure in participatory redevelopment is substantially designated and influenced by the institutional framework, which is dominantly controlled by the state. The institutions decide the distribution and redistribution of structural resources among the involved stakeholders, in terms of exogenous influences. The structural resources decide the actor's position in power structure. The accumulation of allocative and authoritative resources by actors cannot change their positions;
42. The actor's position further determines the effect of the adopted rules. Rules adopted from actors in a higher position to actors in a lower position will generally take effect. If there is no interest conflict between them, the implementation process will be smooth. If there is interest conflict, the exertion of the rules will require high cost. Rules adopted from actors in a lower position to actors in a higher position will come into force only when the rules serve to meet the latter's requirement. High frequency of the rule only enhances the rule per se, but not necessarily realize the aim of the actions;
43. Actors are authorized certain power to make a difference. Their interactions, as a bottom-up force, are able to produce and reproduce the power relations and influence the outcome of the participatory redevelopment. However, their power is continuously regulated and constrained by state power. Their interactions fail to overturn the power structure, which is primarily shaped by the institutional arrangement.

8.2 Conclusion

8.2.1 Conclusion

The participatory reform has advanced urban redevelopment and forged a new development trajectory in Chinese cities since the early 2000s (Li et al., 2021; Li & Liu, 2022). It has reshaped the political structure of urban governance and profoundly influenced power relations among key stakeholders, including the local government, developers, affected residents and the newly emerged real estate agents (X. Li et al., 2019).

To mitigate social resistance and facilitate urban redevelopment, the central government introduced several critical participatory redevelopment policies, initiating a process of power decentralization. The Property Rights Law empowers property owners to contest unfair compensation, while public participation mechanisms institutionalize residents' rights to participate in decision-making. The new land transfer mechanism grants development rights to developers and self-redevelopment rights to existing land users. The abolition of the demolition permits further restricts direct governmental intervention in local redevelopment projects. Collectively, these mechanisms have constrained local government power while empowering residents and developers, thereby reshaping the stakeholders' power structure. Evidence suggests that these new power dynamics effectively reduce social mobilization and facilitate urban village redevelopment (e.g. Chung, 2013; Li & Li, 2011; Zhou, 2014).

However, the empowerment of residents has also introduced substantial challenges to residential redevelopment. The increasing involvement of residents introduces heterogeneous interests and fragmented property rights, complicating the redevelopment process (Hu & Li, 2025; Li et al., 2024). Unlike urban villages, residential redevelopment lacks collective associations to internalize the task of land assembly. Consequently, the local government and developers must negotiate with a large number of households, escalating transaction costs. Equipped with housing expropriation power, the local government can mitigate uncooperative residents' resistance, whereas developers without governmental endorsement are unable to counteract residents' influence, resulting in speculative behaviours and escalating compensation standards. These dynamics negatively affect housing affordability, increase development density and strain infrastructure, leading to higher social costs (He et al., 2019; Shenzhen Urban Planning Research Centre, 2015).

In response to poor performance of market-oriented redevelopment projects, local governments launched a countermovement to reassert control. While experimenting with market mechanisms, the Shenzhen municipal government maintained a government-led approach to sustain intervention. When market-oriented projects underperformed or generated social issues, the government abruptly replaced market mechanisms with government-led strategies. Through these policy adjustments, the local state reasserts control over urban development and strengthens its role in participatory redevelopment.

Contradictions between central and local forces in participatory reform result in an uneven and multidirectional reconfiguration of power relations. Although central policies constrain local government power, the latter continues to dominate decision-making. Rather than relying on forceful interventions, the local government increasingly employs legislative instruments, such as policies and regulations, to assert control, often cloaked under the guise of public interest protection. The empowerment and power deployment of other actors, including residents and market sectors, remain substantially regulated by the local government.

Consequently, power relations in participatory redevelopment are hierarchical in China. Institutional arrangements determine actor hierarchies, primarily reflecting state priorities. Although interactions among actors' influence power dynamics, they are continuously constrained by state authority, which functions as a political oligarchy (Purcell, 2016). No actor, including residents, developers or real estate agents, can overturn the state's jurisdictional hierarchy. These hierarchical power relations are neither collaborative nor inclusive; rather, they are chaotic and conflictual. The local government often establishes policies with minimal consultation, adjusting strategies in response to poor performance through policy replacement, reversal or abolition, resulting in structural instability. Residents must continuously adapt to these changes, creating heterogeneity in values and preferences, which further complicates participatory redevelopment. Structural instability and stakeholder heterogeneity impede the formation of inclusive governance while intensifying conflicts, contradictions and inequalities.

This research thus concludes that the restructuring of power relations has produced an authoritarian participatory governance system in China. Public participation mechanisms have institutionalized residents' involvement, allowing them to negotiate for fair compensation, while the state increasingly considers public welfare in its interventions. However, these measures primarily serve to pacify resistance and facilitate redevelopment rather than to restore the egalitarian planning system of the socialist period. In practice, other actors must cultivate stable, collaborative relationships with the state to achieve their objectives. The state prioritizes its dominance and suppresses social mobilization when necessary, highlighting the authoritarian character of participatory governance. Institutional reforms have established mechanisms for participatory governance to reduce social resistance and facilitate redevelopment, but they primarily enhance state control over urban land and development rather than expanding civil rights or empowering residents to shape future cities.

8.2.2 Transferability to Other Contexts

The findings of this research have potential relevance for other Chinese cities undergoing urban redevelopment, particularly those with strong-state governance structures. Similar hierarchical power relations and state-directed participatory mechanisms are likely to be observed, although local variations in policy implementation, land tenure arrangements and community attributes may influence outcomes.

Beyond China, the study offers insights for urban governance in non-Chinese settings with strong-state characteristics, where institutional arrangements and state authority similarly dominate stakeholder interactions. In such contexts, participatory mechanisms may exist, but the state's strategic control and hierarchical power structures are likely to shape the effectiveness and inclusivity of stakeholder engagement.

8.3 Research Contributions

This thesis contributes to existing studies on Chinese urban (re)development both theoretically and empirically. Theoretically, it develops a power arena framework to examine the impacts of institutional constraints as structural properties, the power configurations of involved actors as agent interactions, and the reciprocal interplay between structure and agents. Empirically, this study provides a systematic and comparative analysis of three patterns of power relations in participatory redevelopment in China. The following sections outline the theoretical and empirical contributions and demonstrate how they address existing research gaps.

8.3.1 Theoretical Contributions

This research develops a novel analytical framework to enhance the investigation of power relations in urban development, addressing the theoretical gaps identified in Chap. 1.

Firstly, the framework simultaneously considers the impacts of institutional forces and local agent power dynamics on power relations. Drawing on Giddens' structure-agent perspective, it posits that structural arrangements establish pre-existing power configurations as exogenous constraints on actors' behaviours. Within these constraints, agents deploy power to interact and advance public agendas. The reciprocal interaction between structural arrangements and agent dynamics produces and reproduces power relations, which in turn shape outcomes.

Secondly, the framework integrates practical analytical tools from institutionalist approaches to theorize structured interactions among stakeholders. The concept of rules is introduced to analyse the functions and effectiveness of actors' actions, facilitating interpretation and comparison of strategies within power relations studies. To further clarify the decision-making process, the framework dissects it into relational sections, or "power arenas", where assembly, disassembly and reassembly of these arenas illustrate the ongoing dynamics of governance.

Thirdly, the framework incorporates power resources as an analytical instrument to identify actors' power configurations and positions. Power resources explain how power derives from structural properties, material assets or authority over others, illustrating how stakeholders are empowered to participate in decision-making. By combining power resources with rules, the framework moves beyond descriptive narratives and provides a structured approach to analysing power relations in urban redevelopment.

8.3.2 Empirical Contributions

This research provides four key empirical contributions that address gaps in studies of participatory redevelopment in China. First, while current studies often attribute transformations in power relations to a singular policy, such as the *San Jiu Gai Zao*, this research treats participatory redevelopment as a comprehensive institutional reform, encompassing property rights reform, constraints on rural land conversion and public participation mechanisms. These institutional changes have generated substantial shifts in urban redevelopment objectives, stakeholder roles, procedures and governance mechanisms.

Second, the study focuses on a unique and timely case city—Shenzhen. Confronted with acute land scarcity, Shenzhen is a prominent policy-testing ground and the first city to promulgate and implement participatory redevelopment policies. Aligned with central directives, the municipal government established local urban redevelopment policies that significantly restructured power deployment among key stakeholders. By addressing the previously limited research on participatory redevelopment in Shenzhen, this thesis fills an important empirical gap and provides a valuable reference for other Chinese cities.

Third, the research investigates residential redevelopment, where property rights are clearly defined but fragmented across individual households. Existing studies predominantly focus on urban village redevelopment, where land is collectively owned. This study reveals distinct stakeholders and power configurations in residential redevelopment, highlighting important differences from urban village contexts.

Fourth, the thesis adopts a multi-case study strategy to conduct a comprehensive and systematic investigation of power relations in participatory redevelopment. Three cases are selected, sharing similar property rights and community attributes but differing in redevelopment mechanisms: government-led, developer-led and owner-led approaches. These cases not only illustrate three distinct redevelopment modes but also trace the evolution of residential redevelopment mechanisms, providing a holistic picture of participatory governance processes in Shenzhen.

8.4 Policy Suggestions

8.4.1 Suggestions for Urban Redevelopment Policies

Government-Led Mechanism

The government-led approach succeeds in facilitating the implementation of residential redevelopment; however, it still manifests some evident cons which need to be addressed. The local public office should be decentralized with more decision-making power to negotiate with residents to revise compensation terms. This can assure the agent rapidly reacts and responds to the requirement of the residents. The grassroots government institutes, such as work stations, should be formally involved to collaborate with the local office for opinion solicitation and negotiation with residents. The engagement of the higher-level governments (district or municipal governments) should be reduced in the initial negotiation process to improve the efficiency of the decision-making process. In this regard, higher-level governments merely take charge of making complementary regulations to constrain and regulate the details of compensation terms, to supervise the decision-making process and to approve the redevelopment schemes. Furthermore, state power in the government-led mechanism should be restricted by a third party such as NGO or the mass media to avert the infringement of developers and residents' interests. Some complementary policies should be established to improve the transparency of the decision-making process, such as regulations on housing expropriation procedures and measures to address property right ambiguity.

Market-Oriented Mechanism

Several suggestions are provided to deal with the poor performance in market-oriented redevelopment approach. The engagements of several developers in one residential area are arguably responsible for the vicious intra-developer competition, which might be well addressed by the auction of the redevelopment right in the government-led project. By means of competitive bidding, developers compete to bid for better building quality and community facilities as well as lower development density. The winner obtains the redevelopment right in the project to negotiate with the affected residents. On one hand, this measure can prevent the engagement of multi-developers to negatively influence residents' life and increase their expectations. On the other hand, it avoids the over-investment of the developers before the agenda setting to prevent asset specificity. This is beneficial to refrain from developer's opportunistic behaviours.

The local government should promulgate complementary policies to curb opportunistic behaviours of the developers and residents. Sanctions on developers' opportunistic behaviours should be established to prevent developers from taking forceful strategies. Regulations on the roof of compensation terms can restrict the speculative behaviours of the uncooperative residents. The engagement of the local government is necessary for market-led redevelopment approach to balance the power distribution between the residents and the developer. On completion of the agreements from the vast majority of the affected residents, the local government should get engaged to compulsorily acquire the properties from the last opposing residents. These measures are arguably beneficial to the success of the market-led redevelopment approach.

Residential Redevelopment Policies

Considering the property rights are fragmentedly owned by individual households in the dilapidated residential areas, the local government should issue complementary policies to control project scale and limit the number of involved residents before approving the residential redevelopment projects. This contributes to reducing the possibility of heterogeneous interests to improve project efficiency.

The coexistence of multiple residential redevelopment approaches in Shenzhen, on one hand, leads to residents' confusion about residential mechanisms; on the other, it generates unnecessary cross-case comparisons among the residents. Meanwhile, the frequent policy changes resulted in recurrent suspension, abolition and replacement of residential redevelopment approaches, adding to the confusions of all the involved stakeholders. The confusions trigger unregulated behaviours and conflicts, raising transaction costs of urban redevelopment. These deficiencies can attribute to the absence of empirical research and public consultation with developers and property owners in the policymaking process. It is, therefore, crucial to increase the engagement of multiple involved stakeholders in the policy-making process. In addition, the local government should adopt more strategies to publicize policies to other involved stakeholders, such as the developer and the affected residents. To make other actors acquainted with the policies will improve the quality of their participation in the decision-making process.

Policy coherence and coordination between governments at different levels are also significant. After the abolition of demolition permit institution at central level, the local government failed to effectively address the administrative conflicts but established a regulation that created great barriers to land assembly. This directly resulted in poor performance and social issues in the market-led project. The local government, therefore, should effectively respond to the central policy change to improve coordination between central and local policies.

It is definitely insufficient for the central government to set up institutional reform exclusively in land use and planning field. The current fiscal system provides significant incentives for the state apparatus at lower levels to profit from urban development. Furthermore, the evaluation of official's promotion in current political and administrative system is heavily biassed towards achievement related to economic growth and capital accumulation at local level (Zhang, 2002). Both forces generate local government's entrepreneurialism and determine the local government's profit-seeking essence in urban development no matter how sophisticated and disguised the measures are. The reform of fiscal arrangement and administrative system is necessary to achieve public agendas, such as social welfare and justice.

8.4.2 Suggestions for the Residential Redevelopment Governance

The participatory redevelopment policies resulted in conflicts and contradictions among the involved stakeholders in the governing process. This section provides some suggestions to promote inclusive and collaboration relationships in participatory redevelopment. The Shenzhen municipal government demonstrates a dichotomy towards state intervention and market mechanism in urban governance. It resorts to either a liberal market mechanism with the least government involvement or a government-led mechanism with powerful state intervention. When the poor performance of market-oriented mechanism occurs, the local state directly relinquished all the market-oriented efforts by roughly abolishing the market mechanism for potential residential redevelopment projects in 2012. The market is merely regarded as a utilitarian instrument to facilitate urban development, while path dependence of planned economy and socialist mindset unreformed (Wu, 2010). In this regard, it would be more effective if the local government promulgates incentives to encourage more market power to facilitate participatory redevelopment. In the meantime, the government sets up clear rules to regulate the market operation and plays a regulatory role to supervise the interacting process between residents and developers.

Developer's poor self-adjust ability to change from a partner of the local government to an independent market player is observed. They are inclined to follow the old path of residential redevelopment: through making major efforts to develop a close relationship with government authorities, they can make prompt project progress and achieve their financial goal. However, the participatory reform has altered the structure of power distribution for decision-making—the residents commence possessing significant decision-making power. Developers will face significant coordination challenge if they ignore the sufficient communication with the affected residents. Therefore, developers should pay more attention to improve their communication skills to obtain the trust and support from the residents. It is necessary to establish a collaborative relationship with the residents to diminishing the cost of reaching a consensus.

The rudimentary development of the self-organized community organization in China fails to appropriately motivate the affected residents to play an active role in the decision-making process. While residents become increasingly aware of their rights and entitlements to participate in civil affairs, the effective engagement of the community in policymaking and decision-making processes has become increasingly important in China. The state should decentralize more autonomy to the self-organized associations such as property owner committee at local level. These associations will help to build up social bonds and facilitate social mobilization at local level, which deepens the community's awareness and involvement in participatory redevelopment initiatives.

8.5 Research Limitations and Future Research Agenda

Limitations of this research lead to follow-on work and future research agenda in three main perspectives. These limitations, however, do not negatively influence this thesis to achieve its objectives and do not substantially affect the credibility of the main conclusion. Instead, they unfold new spaces and possibilities for future research on urban redevelopment and power relations in participatory redevelopment in China.

The first limitation refers to that this research primarily treats the local government as ONE unified actor, paying few attention to the power dynamics among different governmental institutes and authorities within the local state. Two reasons account for the investigation absence. Firstly, the local government always appeared as an integrated actor in urban redevelopment cases in China (e.g. Zhang et al., 2016). Although some internal disputes occurred in the decision-making process, different departments and institutes displayed the uniform behaviour when they were interacting with other stakeholders, such as the affected residents and developers (e.g. the debate on redevelopment approach in Ludancun). This is because local government departments have relatively little political space to resist the government's decisions and they are affiliated to the later and rooted in local politics (Perlstein & Ortolano, 2015). Treating the local government as an ensemble can benefit to focus on the interactions between the state and other actors, to better achieve the research objective. Secondly, it is conducive to conducting a cross-case comparison when seeing the local government as a unified actor. Because of implementing different redevelopment mechanisms, three cases are related to different departments at different levels, such as municipal and district government institutes, in spite of some overlaps. It provided an excellent basement for a cross-case comparison to consider the local government as a single actor with diverse resources and rules. Whereas, the power relations analysis would be possible to extend to the debate and power dynamics among different authorities within the local state in the next stage of research agenda, since the inner debate is able to demonstrate diverse and specific demands of different departments. Although the contradictions may not eventually generate substantial impacts on the final decision of the local government, it manifests the decision-making process of the local state. The manifestation will provide concrete evidence for policymakers and scholars to make suggestions on urban governance.

This research fails to conduct an empirical study on the experimental of residential redevelopment approach—Shanty Area Improvement but conceptualized it as a Public-Private Partnership (PPP) mechanism, leading to the second limitation. Policies encourage the Shanty Area Improvement to initiate a PPP mechanism to cooperate the state power and market power to facilitate urban redevelopment. However, preliminary data collecting from news articles and social media manifest that the local government plays a substantial role in controlling the decision-making process while the involved developers and the affected residents are merely able to execute minor power to influence the governance process. Therefore, whether it is a PPP or a government-led approach is still in question, which requires future research to examine. The empirical analysis of the new approach is important to manifest the evolutionary process of residential redevelopment in China. How the policy impacts the outcome and the power relations among the involved stakeholders deserves future investigation.

The third limitation is that the evaluations of the redevelopment process and outcome have not been fully developed in this research, primarily with respect to process efficiency and transaction costs. In this research, the analysis of transaction costs was principally conducted based on terms related to actors' opportunistic behaviours and asymmetric power relations, failing to comprehensively investigate the relative hazards (i.e. timing, heterogeneity and asset specificity) that generated transaction costs. This is because the objective of this research is to investigate the power relations among different stakeholders and their impacts on the outcomes rather than to conduct an institutional analysis to examine deficiencies of the participatory redevelopment policies. In this regard, focusing on behavioural dynamics and power configurations in process efficiency analysis assists in an in-depth investigation of power relations and their impacts in participatory redevelopment. In the future, the evaluation of process efficiency in terms of transaction costs can be extended to identify the institutional deficiencies of Shenzhen's participatory redevelopment policies.

References

Chung, H. (2013). The spatial dimension of negotiated power relations and social justice in the redevelopment of villages-in-the-city in China. *Environment and Planning A, 45*(10), 2459–2476. https://doi.org/10.1068/a45416

Han, S. S. (2000). Shanghai between state and market in urban transformation. *Urban Studies, 37*(11), 2091–2112. https://doi.org/10.1080/713707226

Han, S. S., & Wang, Y. (2003). The institutional structure of a property market in inland China: Chongqing. *Urban Studies, 40*(1), 91–112.

He, S., Zhang, M., & Wei, Z. (2019). The state project of crisis management: China's shantytown redevelopment schemes under state-led financialization. *Environment and Planning A: Economy and Space, 52*(3).

Hsing, Y. T. (2010). *The great urban transformation: Politics of land and property in China*. Oxford University Press. https://doi.org/10.1093/acprof:oso/9780199568048.001.0001

Hu, W., & and Li, X. (2025). Variegated state entrepreneurialism in China: Land commodification and housing provision in Shenzhen and Chongqing. *Urban Policy and Research, 43*(3) https://doi.org/10.1080/08111146.2025.2472057

Li, B., & Liu, C. (2018). Emerging selective regimes in a fragmented authoritarian environment: The 'three old redevelopment' policy in Guangzhou, China from 2009 to 2014. *Urban Studies, 55*(7), 1400–1419.

Li, B., Tong, D., Wu, Y., & Li, G. (2019). Government-backed 'laundering of the grey' in upgrading urban village properties: Ningmeng apartment project in Shuiwei Village, Shenzhen, China. *Progress in Planning, 46*, 100436.

Li, L., & Li, X. (2011). Redevelopment of urban villages in Shenzhen, China—An analysis of power relations and urban coalitions. *Habitat International, 35*(3), 426–434. https://doi.org/10.1016/j.habitatint.2010.12.001

Li, X., Li, B., & Jiang, W. (2024). State-led versus market-led: How institutional arrangements impact collaborative governance in participatory urban regeneration in China. *Habitat International, 150*, 103134. https://doi.org/10.1016/j.habitatint.2024.103134

Li, X., & Liu, T. (2022). Nomocracy versus teleocracy: Comparing participatory urban redevelopment in Shenzhen and Kashgar. *Journal of Urban Affairs, 46*(2), 298–318. https://doi.org/10.1080/07352166.2022.2053779

Li, X., Wu, H., & Han, S. S. (2021). Institutional deficiencies in market-led residential redevelopment in Shenzhen, China. *Journal of Planning Education and Research, 44*(3). https://doi.org/10.1177/0739456X211051779

Perlstein, A., & Ortolano, L. (2015). Urban growth in China: Evolution in the role of urban planners. *Journal of Planning Education and Research, 35*(4), 435–443.

Purcell, M. (2016). For democracy: Planning and publics without the state. *Planning Theory, 15*(4), 386–401.

Shenzhen Urban Planning Research Centre. (2015). *Compilation and suggestions for urban consolidation policies in Shenzhen from 2012 to 2015.*

Tian, L., & Yao, Z. (2018). From state-dominant to bottom-up redevelopment: Can institutional change facilitate urban and rural redevelopment in China. *Cities, 76*, 72–83.

Wu, F. (2004). Residential relocation under market-oriented redevelopment: The process and outcomes in urban China. *Geoforum, 35*(4), 453–470.

Wu, F. (2010). How neoliberal is China's reform? The origins of change during transition. *Eurasian Geography and Economics, 51*(5), 619–631.

Zhai, B., & Ng, M. K. (2013). Urban regeneration and social capital in China: A case study of the drum tower Muslim District in Xi'an. *Cities, 35*, 14–25.

Zhang, L., Chen, J., & Tochen, R. M. (2016). Shifts in governance modes in urban redevelopment: A case study of Beijing's Jiuxianqiao area. *Cities, 53*, 61–69. https://doi.org/10.1016/j.cities.2016.01.001

Zhang, T. (2002). Urban development and a socialist pro-growth coalition in Shanghai. *Urban Affairs Review, 37*(4), 475–499. https://doi.org/10.1177/10780870222185432

Zhou, Z. (2014). Towards collaborative approach? Investigating the regeneration of urban village in Guangzhou, China. *Habitat International, 44*, 297–305.

Appendix

Observation Protocol

This study employed non-participant observation as a complementary qualitative method to document interaction patterns, decision-making processes and power dynamics among key stakeholders in the redevelopment process. Observations were conducted during on-site visits, community meetings and informal interactions related to the redevelopment project. A semi-structured observation protocol was adopted, with particular attention to the deployment of power resources, the asymmetry of influence, and shifts in relational power across different stages of the redevelopment process. Field notes were recorded contemporaneously and expanded immediately after each observation session to ensure accuracy and reflexivity. The observational data were subsequently triangulated with interview transcripts and documentary materials to enhance the credibility of the findings.

Anonymization and Ethics Statement

This research was conducted in accordance with established ethical standards for social science research. All participants were informed of the purpose of the study and their voluntary participation prior to data collection. To protect confidentiality, the identities of individuals, organizations and specific project actors have been anonymized. Pseudonyms are used for interviewees, and identifying details have been removed or modified where necessary to prevent indirect identification. Given the political and institutional sensitivity of urban redevelopment in China, particular care was taken to ensure that no statements could be traced back to specific individuals or positions. All data were securely stored and used solely for academic purposes.

X. Li, *Participatory Redevelopment as Contested Arenas in China*,
https://doi.org/10.1007/978-981-95-9394-1

Interview Protocol

This interview protocol is designed to investigate power relations in participatory urban redevelopment in China. The questions focus on actors' roles, power resources, strategies, institutional constraints and interactions across different stages of the redevelopment process.

(1) Local Government

(Municipal/District government, Urban Redevelopment Bureau, Planning and Housing Departments).

1.1 Policymaking and Institutional Design

- What are the main objectives of participatory urban redevelopment policies?
- What problems were these policies intended to address?
- How are the roles and responsibilities of different stakeholders (residents, developers, intermediaries) defined in policy design?
- How does the government determine the boundaries of residents' participation in decision-making?
- How is "self-redevelopment power" of property owners understood and institutionalized in current policies?
- Are there mismatches between policy design and local implementation conditions? If so, in what aspects?

1.2 Policy Implementation and Governance

- How does the local government intervene in redevelopment projects during implementation (e.g. approval, coordination, supervision)?
- How does the government balance efficiency, social stability and public participation?
- What are the main challenges encountered during policy implementation?
- How does the government deal with internal conflicts among residents?
- Are policies implemented flexibly in practice? Can you provide examples of policy adjustment or reversal?

1.3 Power Relations and Reflection

- How do you view the continued dominance of the local government in participatory redevelopment?
- Do current policies promote collaborative governance or reinforce hierarchical control?
- Have participatory policies produced uneven outcomes across different communities? Why?
- What are the major limitations of participatory redevelopment in terms of fairness and justice?
- If policies were to be revised, which institutional aspects should be prioritized?

(2) Developers

(Real estate developers involved in participatory redevelopment).

2.1 Roles and Power Resources

- What role do developers play in participatory redevelopment projects?
- How does participatory redevelopment differ from traditional government-dominated redevelopment from a developer's perspective?
- What types of resources do developers mainly rely on (financial, technical, relational)?

2.2 Power Deployment and Negotiation

- At which stages of redevelopment do developers have decision-making power?
- In which aspects is developers' power constrained by the local government?
- How do developers negotiate with the government during redevelopment?
- What strategies are adopted when negotiating with residents?
- Has residents' participation changed developers' implementation strategies?

2.3 Constraints and Risks

- What are the major institutional barriers to participatory redevelopment?
- How does strong governmental intervention affect project feasibility?
- Does residents' collective action increase uncertainty or provide opportunities?
- Do you perceive a decline in developers' power under participatory redevelopment? Why?

2.4 Evaluation and Reflection

- How do you evaluate participatory redevelopment in terms of efficiency, fairness and risk distribution?
- In which aspects would developers prefer greater autonomy?
- Do you believe participatory redevelopment will become the dominant model? Under what conditions?

(3) Affected Residents/Property Owners

3.1 Participation and Rights Awareness

- How did you learn about the redevelopment project and participatory policies?
- What rights do you believe you have in the redevelopment process?
- Which decisions do you think residents can influence, and which are controlled by the government?

3.2 Power Deployment and Collective Action

- How do residents express their demands (meetings, voting, negotiation, intermediaries)?
- Are there resident representatives or organizations? What roles do they play?

- Do differences among residents (information, income, connections) affect participation outcomes?
- How do residents negotiate with the government or developers?

3.3 Constraints on Participation

- Are there stages where residents' participation is restricted? In what ways?
- When residents' demands conflict with government objectives, whose decision prevails?
- How effective is "self-redevelopment power" in practice?

3.4 Perceptions of Justice

- Do you consider the redevelopment outcome fair? Why?
- Does redevelopment affect different social groups differently?
- If given another opportunity, where would you like to have greater influence?

(4) Planners and Technical Professionals

(Planning institutes, design firms, consultants).

4.1 Professional Role and Technical Power

- What is your role in the redevelopment project?
- To what extent do technical plans influence final decisions?
- Have you experienced tensions between technical rationality and administrative authority?

4.2 Mediation and Coordination

- Do planners act as mediators between the government, residents and developers?
- Whose opinions ultimately shape planning outcomes?
- Are planning and technical tools used to legitimize predetermined decisions?

4.3 Institutional Constraints and Reflection

Do existing institutional arrangements limit professional judgement?
How substantive is planners' participation in participatory redevelopment?
How could planners better support collaborative governance?

(5) Real Estate Agents and Other Intermediaries

(Real estate agents and community organizations).

5.1 Role and Entry into the Process

- How did you become involved in the redevelopment project?
- How do you define your role?
- Whose interests do you primarily represent?

5.2 Power Resources and Constraints

- What resources do you rely on to influence decisions (information, organization, networks)?
- Are your opinions considered in key decision-making processes?
- What factors constrain your influence?

5.3 Marginalization and Reflection

- Do you feel marginalized in the decision-making process?
- How does the limited effectiveness of self-redevelopment power affect your role?
- What positive functions could intermediaries perform under a more open institutional framework?

(6) Concluding Questions (for All Interviewees)

- How would you characterize power relations in participatory redevelopment?
- Does the current system prioritize efficiency, stability or fairness?
- Has participatory redevelopment fundamentally altered traditional power structures?
- What institutional reforms are most urgently needed?

Interviewers List

Date	Amount	Participants Number	Role	Work Unit	Name	Time	Case
27/11/2016	1	41	Public servant	Urban Planning and Land Resources Committee in Long Gang	WHZ	20 min	
28/11/2016	1		Public servant	Luohu Urban Renewal Authority	LXY	32 min	Mutoulong
29/11/2016	1		Public servant, designer	Shenzhen Urban Planning and Design Institute	LQ	10 min	Mutoulong
	1		Developer of Mutoulong	Yitian Real Estate Company	LPL	15 min	Mutoulong

(continued)

(continued)

Date	Amount	Participants Number	Role	Work Unit	Name	Time	Case
1/12/2016	1		Planner	Shenzhen planning and land Development Research Center	ZNQ	50 min	
	1		Planner	Shenzhen planning and land Development Research Center	LCS	75 min	
7/12/2016	1		Agreement-signed owner	/	CL	90 min	Mutoulong
	1		Agreement-signed owner	/	/	15 min	Mutoulong
	1		Agreement-signed owner	/	/	20 min	Mutoulong
	1		Agreement-unsigned owner	/	/	50 min	Mutoulong
8/12/2016	1		Developer of Ludancun	Zhonghai Real Estate Company	LYB	75 min	Ludancun
9/12/2016	1		Planner	Shenzhen planning and land Development Research Center	LCS	60 min	
12/12/2016	1		Agreement-unsigned owner	/	/	20 min	Mutoulong
14/12/2016	1		Developer of Ludancun	Zhonghai Real Estate Company	FXH	60 min	Ludancun
15/12/2016	1		Designer	Shenzhen Urban Planning and Design Institute	WJ	40 min	
20/12/2016	1		Public servant	Shenzhen Urban Renewal Authority	CSH	30 min	Ludancun

(continued)

(continued)

Date	Amount	Participants Number	Role	Work Unit	Name	Time	Case
22/12/2016	1		Designer	Shenzhen planning and land Development Research Center	ZYL	30 min	Mutoulong
28/12/2016	1		Public servant who took in charge of	Shenzhen Housing Security Authority	WDB	30 min	Ludancun
	1		Relocated owner in LG	/	YBX	30 min	
3/01/2017	1		Property owner of Nanhuanun	Nanhuanun property owner committee	WLP	90 min	Nanhuanun
	1		Public servant	Nanhuanun work station	D	40 min	Nanhuanun
	1		Public servant	Ludancun work station	L	30 min	Ludancun
4/01/2017	1		Late cooperative resident	/	/	10 min	Ludancun
5/01/2017	1		Property owner of Nanhuanun	/	W	120 min	Nanhuanun
6/01/2017	1		Public servant	Futian Urban Renewal Authority	YY	50 min	Nanhuanun
8/01/2017	1		Developer of Mutoulong	Yitian Real Estate Company	R	140 min	Mutoulong
9/01/2017	1		Cooperative owner	Vice head of senior resident association	HY	70 min	Ludancun
	1		Manager of Nanhuanun redevelopment	Vanke Real Estate Company	ZJ	40 min	Nanhuanun, Mutoulong
12/01/2017	1		Public servant	Futian Urban Renewal Authority	Y	40 min	Nanhuanun
	1		Public servant	Nanyuan subdistrict office	LZR	60 min	Nanhuanun

(continued)

(continued)

Date	Amount	Participants Number	Role	Work Unit	Name	Time	Case
15/01/2017	1		Cooperative owner	Head of Property owner committee	WXN	40 min	Nanhuanun
16/01/2017	1		Lawyer	Guanghe law office	LFQ	40 min	Mutoulong, Ludancun
17/01/2017	1		Relocated owner in LG	/	YBX	30 min	Ludancun
18/01/2017	1		Cooperative owner	/	WBM	20 min	Nanhuanun

Glossary

Government-led redevelopment 政府主导更新
Developer-led redevelopment 开发商主导更新
Owner-led redevelopment 业主主导更新
Self-redevelopment 自主更新
Rules 规则
Power Resources 权力资源
Power Arena 权力竞技场
Transaction Costs 交易成本
San Jiu Gai Zao 三旧改造
Cun Liang Gui Hua 存量规划
Zeng Liang Tu Di 增量土地
Peng Hu Qu Gai Zao 棚户区改造
Tu Di Zheng Bei 土地整备
Dilapidated Residential Area (DRA) 老旧住区
Demolition Compensation Ratio (DCR) 拆赔比
Floor Area Ratio (FAR) 容积率
Village Joint Venture Enterprise (VJVE) 村集体合资企业
Property Owner Committee (POC) 业主委员会
Senior Resident Association (SRA) 老年居民协会
Guannei 关内
Under-the-table deal 私下交易
Work Unit 单位

X. Li, *Participatory Redevelopment as Contested Arenas in China*,
https://doi.org/10.1007/978-981-95-9394-1

Bibliography

Fischer, E., & Qaim, M. (2012). Linking smallholders to markets: Determinants and impacts of farmer collective action in Kenya. *World Development, 40*(6), 1255–1268. https://doi.org/10.1016/j.worlddev.2011.11.018

He, S., Zhang, M., & Wei, Z. (2019). The state project of crisis management: China's Shantytown Redevelopment Schemes under state-led financialization. *Environment and Planning A: Economy and Space*, 0308518X19882427

Li, X., Han, S. S., & Wu, H. (2019). Urban consolidation, power relations, and dilapidated residential redevelopment in Mutoulong, Shenzhen, China. *Urban Studies, 56*(13), 2802–2819. https://doi.org/10.1177/0042098018799950

Li, Y., & Feng, S. (2016, Dec–12). Why did the tough uncooperative residents sign agreement with the developer in Mutoulong? *Southern Metropolis Daily*. http://www.oeeee.com/html/201612/12/437508.html

Pred, A. (1985). The social becomes the spatial, the spatial becomes the social: enclosures, social change and the becoming of places in Skåne. In *Social relations and spatial structures* (pp. 337–365). Springer.

Schoon, S., & Altrock, U. (2014). Conceded informality. Scopes of informal urban restructuring in the Pearl River Delta. *Habitat International, 43*, 214–220.

The residents in Nanhuacun looking forward to redevelopment for more than 17 years. (2014, May–10). *Southern Metropolis Daily*.

The State Council. (2015). *Opinions of the state council on further improving the redevelopment of Shanty areas and Urban and Rural decrepit housings and the construction of related infrastructure.*

Tian, C. (2016, Jan–14). Dozens of housing were identified as decrepit housing but the redevelopment is in stall. *Shenzhen News*. http://dc.sznews.com/content/2016-01/14/content_12735295.htm

Wu, W., Zhao, S., Zhu, C., & Jiang, J. (2015). A comparative study of urban expansion in Beijing, Tianjin and Shijiazhuang over the past three decades. *Landscape and Urban Planning, 134*, 93–106. https://doi.org/10.1016/j.landurbplan.2014.10.010

Zhang, T. W. (2000). Land market forces and government's role in sprawl - The case of China. *Cities, 17*(2), 123–135. https://doi.org/10.1016/S0264-2751(00)00007-X

X. Li, *Participatory Redevelopment as Contested Arenas in China*,
https://doi.org/10.1007/978-981-95-9394-1

Index

X. Li, *Participatory Redevelopment as Contested Arenas in China*,
https://doi.org/10.1007/978-981-95-9394-1